Perspektiven der Mathematikdidaktik

Herausgegeben von
G. Kaiser, Hamburg, Deutschland

In der Reihe werden Arbeiten zu aktuellen didaktischen Ansätzen zum Lehren und Lernen von Mathematik publiziert, die diese Felder empirisch untersuchen, qualitativ oder quantitativ orientiert. Die Publikationen sollen daher auch Antworten zu drängenden Fragen der Mathematikdidaktik und zu offenen Problemfeldern wie der Wirksamkeit der Lehrerausbildung oder der Implementierung von Innovationen im Mathematikunterricht anbieten. Damit leistet die Reihe einen Beitrag zur empirischen Fundierung der Mathematikdidaktik und zu sich daraus ergebenden Forschungsperspektiven.

Herausgegeben von
Prof. Dr. Gabriele Kaiser
Universität Hamburg

Sasha Wang

Discourse Perspective of Geometric Thoughts

With a foreword by Anna Sfard, University of Haifa

 **Springer** Spektrum

Sasha Wang
Boise State University
Idaho, USA

Perspektiven der Mathematikdidaktik
ISBN 978-3-658-12804-3   ISBN 978-3-658-12805-0 (eBook)
DOI 10.1007/978-3-658-12805-0

Library of Congress Control Number: 2016933534

Springer Spektrum

Printed on acid-free paper

This Springer Spektrum imprint is published by Springer Nature
The registered company is Springer Fachmedien Wiesbaden GmbH

Foreword

What does geometry, the science of space and shape, have to do with talking, the art of juggling words? On the face of it, not much. Indeed, geometry, at least as it is learned in schools, seems to be occurring mainly in images, and images, they say, speak for themselves. At a closer look, however, geometry begins where the visual means alone can no longer tell a story. This is what happens when *relations* between shapes, rather than the shapes as such, become the protagonists of the narrative. Through words, geometric figures stop being a collection of lonely individuals and start showing hitherto unsuspected family ties. Indeed, it is because of the common elements detectable within *definitions* of figures rather than of the figures themselves that we cluster differently looking shapes into families. Indeed, the family resemblance of geometric figures resides in verbal patterns displayed by the stories we tell about them. For instance, we turn the term *rectangle* into the surname of the square because the proposition "this is a right-angle parallelogram" that constitutes the definition of rectangle appears also in the verbal descriptions of the square. Drawing on this example, one can say that geometry is the art of transforming *visual* patterns into propositions and then translating the resulting *verbal* patterns back into stories about figures. As such, geometry has all to do with words!

The realization of this fact is the point of departure for Dr. Wang's research presented in this volume. As signaled already by its title, *Discourse Perspective of Geometric Thoughts*, her project draws on the assumption that words, rather than being mere descriptors of geometric objects, are the very fabric of which geometry is made. As the author shows in this book, the non-standard foundational tenet according to which geometry is a discourse brings a new lease of life to the time-honored and densely populated field of study. To understand how this foundational transformation changes our thinking on learning on the universe of geometric figures, one has to keep in mind that discourses, that is, well-defined forms of communication, are interpersonal affair: they originate between people and develop in human interactions. This fact locates the resulting theory within the socio-cultural paradigm. Here, learning is seen as an inherently collective endeavor, regulated by social forces even in those cases in which the learner does her work in a temporary isolation from other people. This, in turn, implies that the student cannot be seen as the sole author of either her own persistent failure or her lasting success. This vision is what sets Dr. Wang's research apart from the previous work on the development of geometric thinking, most of which has been inspired by Piaget's cognitivist vision of individual development and learning.

In particular, the conceptualization of geometric thinking presented in this book is quite different from the arguably most popular theory proposed nearly sixty years ago by Dina and Pierre van Hiele. And yet, for all the difference between Dr. Wang's socio-cultural and van Hiele's cognitivist approaches, this book makes it clear that the time honored model does not lose its appeal in the eyes the discursive researcher. In fact, it seems only natural to redefine the famous van Hiele levels of geometric thinking in discursive terms. One can speculate that the van Hieles themselves, having repeatedly stressed that linguistic changes are one of the main characteristics of the transition from one level to another, might have been pleased with such discursive recasting. To me, as to the author of this book, it is quite clear that the discursive redo of their model is indispensable if we wish to go beyond the mere statement about the existence of linguistic transformations and want to see what actually happens to the students' language as she climbs the five-step ladder.

Discursive refurbishing of van Hiele model and then refining it with the resulting insights is exactly what Dr. Wang does in her research. Her discursive reworking, far from being a mere translation into a different language, should rather be thought of as an act of grafting time-honored ideas onto a new line of thought. This kind of crossbreeding makes the resulting structure as different from the original one as the hybrid plant obtained by grafting is different from its source. The new model of the development of geometric thinking that emerges from these efforts seems to amplify the strengths of the old one, while also adding a number of new ones.

One of the special advantages of the scheme proposed by Dr. Wang is its operationality, the property without which research cannot be truly rigorous and convincing. Unlike in many previous cases, in which the concern about scientific precision led to extreme reductionism, this important gain does not come here at the expense of the researcher's ability to capture geometrical thinking and learning in all their complexity. On the contrary, Dr. Wang's method makes it possible to attend to most subtle aspects of discursive activity of the learner and to make distinctions that are too subtle to capture with the less sensitive cognitivist tools. The high-resolution picture of the process of learning that results from this kind of analysis embeds previous research findings in a new context, thus changing our understanding of well-known phenomena. One such change occurs in this book when students' evolving discourse about the relations of congruence and similarity between geometric figures is examined in a close-up.

This book, with its novel vision of what geometry and with the accompanying illustration of how discursive reconceptualization opens new vistas on geometric thinking, is an invitation to a whole new line of studies. As such, it should be read by those researchers and practitioners who believe that only by revising our own thinking about geometry can we ever hope to improve the teaching and learning of this famously problematic school subject.

Anna Sfard
University of Haifa
November 2015

Preface

In mathematics education, researchers bring in interdisciplinary expertise in language and discourse through the lenses of learning sciences to study the relationship between language, discourse and mathematical thinking, as well as the significance of linguistic or semiotic elements of language and discourse in the development of mathematical thinking. Learning mathematics is a challenge for students because of its abstract nature. Moreover, in mathematics, words such as "function" "parallelogram," and "fraction" have their own expressions and structures to signify complex mathematical concepts, and this adds more of a challenge not only to the understanding of mathematics but also to the communication about mathematics. This study sets out to explore primary pre-service teachers' geometric thinking and ways they communicate their thinking through discourse, and to investigate the characteristics of geometric discourse and its development.

Drawing on a sociocultural with participationist view, Sfard's (2008) *Thinking as Communicating: Human Development, the Growth of Discourses, and Mathematizing*, considers learning as moving towards a more sophisticated mathematical discourse through participation and conceptualizes mathematics as a type of discourse. Sfard also offers an analytical framework that mathematical discourse is distinguishable by its word use, visual mediators, routines, and narratives. In learning and teaching geometry, the van Hieles (1959/1985) view learning as moving towards a higher level of thinking and provide a model that describes five levels of geometric thinking, as visual, descriptive, theoretical, formal logic, and rigor, through Levels 1 to 5, respectively. This study has undertaken a translation of the van Hiele model into discursive terms, called *the Development of Geometric Discourse*. Building on Sfard's communicational approach and empirical data, the study also attempts to address the weakness of the van Hiele model, by proving subtle differences in geometric discourse within each van Hiele level.

The inspiration for the book is to recognize the need for a greater focus on the relationship between mathematical thinking, discourse, and communication to support student learning in today's classrooms.

Acknowledgments

This book is an expansion of my dissertation, *The van Hiele Theory through the Discursive Lens: Prospective Teachers' Geometric Discourses*, completed at the Michigan State University, United States. I would like to acknowledge five individuals from Michigan State University who provided guidance, read the original draft of the dissertation, and gave exceedingly helpful suggestions for its improvement. Glenda Lappan and Anna Sfard served as my doctoral co-advisors. Robert Floden, Gerald Ruden and Sharon Senk (emerita) served on my dissertation committee. I also would like to thank Zalman Usiskin from the University of Chicago (emerita), who provided extensive comments and suggestions for an earlier version of my dissertation. Finally, I wish to acknowledge two other individuals: Jinfa Cai from University of Delaware, whose continual encouragement throughout the writing of the book from the beginning to the end I greatly appreciated, and Gabriele Kaiser from the University of Hamburg, Germany, who provided this great opportunity of publishing this book by Springer.

I would be deeply remiss if I did not specifically acknowledge a group of college students who participated in my dissertation, as they were my primary source of knowledge. I learned tremendously from their geometric thinking and the way they communicated their thinking to me. It is their participation and generous support that made it possible to complete the intended study.

I also want to extend my gratitude to Ms. Tina Starnes and Mr. John Shepard for their help in editing and formatting the manuscript. I am indebted to all the people who supported me in pursuing this dream.

Sasha Wang
Boise State University
October 2015

Table of Contents

Figures

Tables

Nomenclature

CCSSM	Common Core State Standards for Mathematics
NCTM	National Council of Teachers of Mathematics
NGACBP	National Governors Association Center for Best Practices
PSSM	Principles and Standards for School Mathematics
CDASSG	Cognitive Development and Achievement in Secondary School Geometry
GLEs	Grade Level Expectations

1 Chapter One: Introduction

In a research report prepared for the U.S. Department of Education Wilson, Floden and Ferrini-Mundy (2001) reported that research shows a positive connection between teachers' preparation in their subject matter and their performance and impact in the classroom, and found "current results of subject matter preparation are disappointing" (p.35). Darken (2007) also pointed out that "the weak mathematical preparation of many elementary and middle school (K-8) teachers is one of the most serious problems afflicting American education" (p.20). The current document of the Common Core State Standards for Mathematics (CCSSM) sets the standards for mathematics practices describing varieties of mathematical proficiencies in reasoning abstractly, critiquing the reasoning of others and critical thinking (NGACBP, 2010). These concerns and recommendations suggest that a teachers' preparation program needs to emphasize both mathematics content knowledge and pedagogical content knowledge for teaching. Knowing mathematics for teaching involves knowledge of mathematical ideas, mathematics reasoning skills, communication skills, fluency with examples and terms, and thoughtfulness about the nature of mathematical proficiency.

In mathematics, geometry, which is considered a tool for under-standing and interacting with the space in which we live, is perhaps the most intuitive, concrete and reality-linked part of mathematics. It is in the language of geometry that the visual structure of our physical world is described and communicated between individuals and this language of geometry helps students to reason deductively and to think inter-dependently. Today, the language of geometry is used without its structure and grammar, and thus is still a foreign language to many teachers (Pimm, 1987; Usiskin, 1996). A recent report documented primary school teachers received little training in geometry in their teacher preparation programs (Banilower, Smith, Weiss, Malzahn, Campell and Weis, 2013).

The National Council of Teachers of Mathematics (NCTM, 2000) *Principles and Standards for School Mathematics (PSSM)* recommends that students "analyze characteristics and properties of two- and three dimensional geometric shapes and develop mathematical arguments about geometric relationships" (p.41). For instance, in the Geometry Standards for students in grades 3 to 5, it is recommended that all students identify, compare and analyze polygons, develop vocabulary to describe their attributes, classify polygons according to their properties, and develop definitions of classes of shapes. It is well documented that identifying, describing, comparing and classifying geometric shapes, and reasoning with shapes and their attributes are key concepts in school geometry across Kindergarten to Grade 5 (NGACBP, 2010). Because students are expected to learn about geometrical concepts and attributes, as well as relationships between them, it is important for future teachers to know and be comfortable with the concepts and the technical language in geometry relating to these concepts.

Research on students' learning of geometry has addressed the complexity and difficulty in learning school geometry at all levels, as well as other educational and psychological concerns. Recently, the mathematics education community became more aware of the importance of teacher and student interaction in the classroom and how this interaction influences the effectiveness of teaching and learning. The notion of mathematics as discourse and students as being apprenticed into particular ways of *doing mathematics* in particular discursive contexts is now gaining prominence in mathematics education research. This phenomenon prompted the call for a study of pre-service teachers' knowledge in geometry and of their learning of geometry.

While previous work sheds light on pre-service teachers' knowledge and thinking in geometry, it has not explored how examination of these pre-service teachers' geometric discourse could help in learning more about their levels of geometric thinking. This study draws on the sociocultural view of learning, is influenced by the discursive nature of van Hiele model, and of discourse analysis in the form of a discursive framework (Sfard, 2008). It seeks to examine pre-service teachers' knowledge in geometry and to investigate their ways of communicating geometric thinking. This study revisits the van Hiele model with careful examination of key features of mathematical discourse at each level. These features include (1) use of mathematical words, (2) use of visual mediators in the form of geometric figures and their parts, and symbolic artifacts created for the purpose of communicating about geometry, (3) endorsed narratives such as mathematical propositions, axioms and definitions, and (4) mathematical routines as repetitive patterns and course of actions that can be observed. The discursive framework provides a new theoretical approach on mathematical thinking—a form of communication through the analysis of mathematical discourse. Thus, one of the significances of the study is to advance our knowledge about learning—changes in thinking, discourse and social interactions.

Many primary school teachers lack professional preparation in mathematics (Banilower et al, 2013). However, recent calls for improvements in mathematics classrooms instruction have focused on promoting deep scientific reasoning and making connections with core ideas in mathematics. Teachers are facing challenges in making this vision a reality in their classrooms. To support effective teaching and learning, there is an urgent need to build a professional knowledge base for teachers and educators identifying effective classroom practices. Sfard's communicational approach to mathematics offers a kind of learning experience wherein learning mathematics through a process of knowledge construction requires learners to actively engage in social interactions. Social interaction provides opportunities to use others as resources, to share ideas, and to participate in high-quality discussions in which learners learn how to communicate those ideas. Thus, the discursive framework also has a practical implication for improving instructional practices for teachers by creating a discussion-based, dialogue-oriented learning environment.

Given the increased interest in social dimensions of mathematical thinking and learning and that Sfard's (2008) communicational approach to mathematical discourse is still new to many researchers and teacher educators, this study is motived to explore the framework with the following questions:

1. At what van Hiele levels do pre-service primary teachers operate?
2. What are the characteristics of pre-service primary teachers' geometric discourse at each van Hiele level?
3. What additional information (if any) does Sfard's discursive framework provide with regarding to pre-service primary teachers' levels of geometric thinking?

Chapter 2 sets the position of this study among other studies addressing the teaching and learning of geometry in mathematics education in general, and studies that examine students' thinking in geometry using the van Hiele model, and studies emphasizing discursive learning. In addition, this chapter introduces Sfard's discursive framework in detail, including a description of each of its four key mathematical features and important phenomenon highlighted in this framework. Chapter 3 describes the methodology of the study, including descriptions of the van Hiele Geometry Test instruments, interview tasks and an outline of the design of the study. Chapter 4 contains the results of the analyses conducted in this study along with interpretations of findings. The findings are reported in two sections: the results of the van Hiele Geometry pre-test and post-test analysis and the results of pre-interview and post-interview analysis. The test results provide initial information of pre-service primary teachers' van Hiele levels of geometric thinking. The interview results reveal details of these pre-service teachers' geometric discourse at each van Hiele level. Finally, Chapter 5 provides a discussion of the results.

2 Chapter 2: Theoretical Basis

This chapter provides a theoretical basis of the study. It has two sections. I first review the relevant literature and then I provide a theoretical framework for the study.

2.1 Review of Relevant Literature

In this literature review, the first section describes the van Hiele model and then summarizes studies guided by the model in investigating the learning of geometry. In addition, this section also summarizes research that addresses the knowledge of mathematics for teaching geometry. The second section describes Sfard's discursive framework and her communicational approach to the issue of learning and nature of mathematical thinking. Included in this section are summaries of discourse studies in the mathematics classrooms and a translation of geometric discourse that aligns the van Hiele model through the discursive lens. The final section raises general research questions in discursive terms.

For the teaching and learning of geometry, the van Hieles developed this influential model to distinguish levels of geometric thinking. In discussing the profound impact of the van Hiele model in mathematics education, Clements (2003) concluded that van Hiele "gave educators and researchers a model that promoted the understanding of important, conceptual based level of thinking... It is also a model of synergistic connections among theory, research, the practice of teaching, and students' thinking and learning" (p.151). To better describe the van Hiele model and how it has been used in the field of mathematics education, the following section provides the historical background and a general description of the theory.

2.1.1 The van Hiele Model

The root of the van Hiele model emanated from the task of improving the teaching of geometry. A Dutch husband and wife, Pierre Marie van Hiele and Dina van Hiele-Geldof, developed "the van Hiele model" in their doctoral dissertations at the University of Utrecht, Netherlands, in 1957. Dina died shortly after completing her dissertation, and Pierre continued to develop and disseminate the theory (van Hiele 1959/1985,1986).

When Pierre and Dina worked at Montessori secondary schools as mathematics teachers, they were very disappointed with "students' low-level knowledge of geometry" (van Hiele, 1959/1985, p.60). On the other hand, they also realized that teachers and students often fail to communicate with each other because they "speak a very different language" (p.61). For example, one of Pierre and Dina's initial observations was that they seemed to speak about geometry in a different way than their students. When Pierre and Dina spoke about a square as a type of rectangle, students were confused because to them a square and a rectangle were quite different. This led Pierre and Dina to consider the existence of various levels of geometric thinking and the possibility that those students and teachers at different levels of thinking may have difficulty communicating with one another. Although Pierre and Dina developed the theory together, their views were quite different. As a result, Pierre's dissertation focused on identifying students' levels of thinking in learning geometry, while Dina's dissertation was more about a teaching experiment designed to investigate how students move from level to level.

The van Hiele model includes five distinct levels that describe students' thought levels in geometry. However, Pierre van Hiele suggested that mathematics educators should focus on the first four van Hiele levels, because those are what teachers have to deal with in school most of the time (van Hiele, 1986). As Pierre van Hiele (1959/1985) described in "the Children's thought and geometry," the five van Hiele levels are as follows (p.62-63): Base Level, figures are judged by their appearance; First Level, figures are bearers of their properties, and they are recognized by their properties but not yet ordered; Second Level, properties are ordered and they are deduced one from another; at this level, definitions of figure come into play but students do not understand the meaning of deduction; Third Level, thinking is concerned with the meaning of deduction, with the converse of a theorem, with axioms, with necessary and sufficient conditions; Fourth Level, thinking is concerned with a variety of axiomatic systems that are non-Euclidean. Geometry is seen in the abstract.

As described in the levels, students' levels of thinking attached to the learning of a particular geometric topic are inductive in nature. At level n, the objects studied are now the statements that were explicitly made at the previous level (level n-1) as well as explicit statements that were only implicit at level n-1. At level n-1, certain geometric objects are studied. Students are able to state some of the relationships explicitly about the objects. Therefore, the objects at level n consist of extensions of the objects at level n-1. One major purpose of distinguishing the levels is to recognize obstacles that are presented to students. For example, when a student who is thinking at level n-1 confronts a problem that requires vocabulary, concepts or thinking at level n, the student is unable to make progress on the problem, with expected consequences such as frustration, anxiety and even anger.

The van Hiele levels have several important properties. First, the levels are discrete and sequential. *Discrete* indicates that the levels are qualitatively different from one another. *Sequential* implies that students pass through the levels in the same order, although varying at different rates, and it is not possible to skip levels. Second, that which was intrinsic at one level becomes extrinsic at the next level. For example, students operating at Level 1 are able to name geometric figures only by their appearance as a "whole"—the properties of a figure remain intrinsic. However, at Level 2, these properties become extrinsic and in fact are the new objects of study. Third, each level has its own language and symbols. The van Hieles believed that "in general, the teacher and the student speak a very different language" (van Hiele, 1986, p. 62). Therefore, teachers and students often have difficulty communicating with one another about geometric concepts. This linguistic challenge can also extend to communicational difficulties between students in a classroom when they are functioning at different thought levels. Finally, instructional methods have a greater influence than either age or grade on a student's progress through the van Hiele levels. That is, a teacher's instructional activities can either foster or impede movement through the levels.

When assigning students to different van Hiele levels, Pierre van Hiele cautioned that it is possible to misjudge a student's level of thinking without careful analysis, because often students memorize or learn patterns in order to accomplish tasks, but do not really understand the underlying concepts. An example is when students recognize corresponding angles by finding the 'F' that is formed by parallel lines and the transversal (see Figure 2.1)

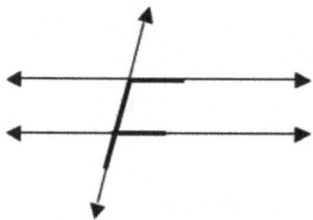

Figure 2.1. Corresponding angles of parallel lines intersected by a transversal.

This technique simplifies the relation between angles and lines. Pierre van Hiele claimed that it could be harmful to students if they only seek a quick result and avoid the "crisis of thinking." In saying "crisis of thinking," he meant the difficulties that students have to transition from one level to a higher level. It is possible for students to derive the answer without recognizing the relationships between the angles in the figure (e.g., supplementary angles, angles at a point, interior angles). Pierre van Hiele warned that these types of "tricks" might actually prevent students from moving to the subsequent level of reasoning (van Hiele,1986, p.42).

The van Hiele model recognizes the importance of language, which plays a significant role in communicating geometric thinking. According to Pierre van Hiele, students' levels of thinking are important not in the sense of the way of their thinking, but in the results of thinking that are revealed in students' speaking and writing. For example, the meaning of a statement like, "This figure is a rhombus" depends on how one argues about it. For a student who is at Basic level, her/his reasoning could be, "This figure has a shape that looks like what I learned to call 'rhombus'." In contrast, if another student has already obtained the first van Hiele level or higher, her/his reasoning could be quite different. The figure that the student refers to is a collection of properties and those properties he/she has learned to call "rhombus" (van Hiele, 1986, p.109). By making the same statement, "This figure is a rhombus," two students could use different reasoning as a result two different levels of thinking. This example of students' responses to a rhombus illustrates how the same statement provided by students can have different meanings because their geometric thinking varies among levels.

2.1.2 Research Guided by van Hiele Model

The van Hiele model has been influential and extensively studied. In this section, the review of the existing literatures focuses on how the model has been used in research in the years since it was developed. Following its incorporation into a new Soviet geometry curriculum, the model was introduced to the United States by the Russian mathematician Izaak Wirszup in a lecture titled "Some Breakthroughs in the Psychology of Learning and Teaching Geometry" at the Closing General Sessions of the National Council of Teachers of Mathematics in 1974 (Wirszup, 1976; van Hiele, 1959/1985; 1986). After the van Hiele levels were translated into English, they were widely used by many researchers in the United States. During the period of 1980-83, the National Science Foundation funded three major investigations using the van Hiele model in the United States: one directed by Burger and Shaughnessy at Oregon State University (1986), another by Fuys, Geddes, and Tischler at Brooklyn College (1988), and a third by Usiskin at the University of Chicago (1982). Burger and Shaughnessy set up a study using clinical interviews to determine the usefulness of van Hiele levels for describing geometric thinking in elementary, middle, and high school grades. Fuys and his colleagues (1988) focused their investigation on geometric thinking in adolescents using instructional models. Usiskin's (1982) project used a large-scale survey to test whether the van Hiele theory applied to the geometric reasoning of students enrolled in secondary geometry courses. These three intensive studies have been widely read, discussed, and cited. After these studies, dozens of other studies using the work of the van Hieles have been conducted in the United States. For instance, Mayberry (1983) used the model to investigate pre-service teachers' geometric thinking, and Senk (1983; 1989) used it to assess high school students' reasoning and proof in geometry.

There is also scarcity of research related to the van Hiele model of reasoning to assess students' geometric thinking. For example, Hoffer (1981) considers the model is integrated with "five skills in geometry" (Hoffer, 1981), whereas Micheal de Villiers (1987) identifies the model as "six geometric thought categories." Gutierrez and Jaime (1998) suggest different processes of reasoning including *recognition*, *definition*, *classification* and *proof* as characteristic of several van Hiele levels. Battista (2009) characterizes the van Hiele model as five levels of reasoning: visual-holistic reasoning (Level 1), descriptive-analytic reasoning (Level 2), relational-inferential reasoning (Level 3), formal deductive proof (Level 4), and rigor (Level 5) (pp. 92-94). These studies led to the conclusion that one cannot consider a level of reasoning as a singular process that is attained (or not) by students, but must be considered as a set of processes.

In the earlier writing of the van Hieles, the van Hiele levels of geometric thinking mainly refer to the classification of quadrilaterals (van Hiele, 1959/1985). At that time, levels were descriptors and they were not labeled by single words (e.g., "Visual" for Level 1, etc.). Almost thirty years later, however, van Hiele (1986) referred to the five levels of thinking as visual, descriptive, theoretical, formal logic, and rigor, and considered such classification to be suitable to a structure of mathematics (p.53).

Over the years, researchers not only used the levels to study students' levels of geometric thinking, but also expanded the area of research from classification of the quadrilaterals to the classification of similar figures, to reasoning and proof, to spatial geometry in three-dimension measurement, etc. In these studies, researchers have proposed various descriptive labels for the van Hiele levels. As the first to name the van Hiele levels, Hoffer (1981) provided his descriptors, "levels of mental development in geometry" (p.13), which label Levels 1 through 5 as recognition, analysis, ordering, deduction, and rigor (p.13-14). Besides these five levels, Hoffer also suggested five basic skills in geometry that are expected at each level. These five skills are visual skills, verbal skills, drawing skills, logical skills, and applied skills. For instance, at Level 1 (recognition), the visual skills only focus on recognizing different figures from a picture, or on recognizing information labeled on a figure. At Level 2 (analysis), visual skills are developed to notice properties of a figure as well as to identify a figure as a part of a larger figure. At Level 3 (ordering), visual skills help to recognize interrelationships between different types of figures and common proper-ties of different types of figures. At Level 4 (deduction), visual skills focus on using information about a figure to deduce more information. Finally, at Level 5 (rigor), visual skills are used to recognize unjustified assumptions made by using figures (p.15). Hoffer's framework suggested that in developing geometric reasoning various geometric skills might be expected at different van Hiele levels of thinking.

Other "level indicators" suggested by Burger and Shaughnessy (1986), describe the five levels as visualization, analysis, informal deduction, formal deduction, and rigor, for Levels 0 through 4, respectively. Using Burger and Shaughnessy's level indicators, Crowley (1987) provided additional examples of level-specific responses (except for Level 5) concerning how students would argue a given shape is a rectangle (p. l5).

Level 1   "It looks like one." Or "Because it looks like a door."
Level 2   "Four sides, closed, two long sides, two shorter sides, opposite sides parallel, four right angles ..."
Level 3   "It is a parallelogram with right angles."
Level 4   "This can be proved if I know this figure is a parallelogram and that one angle is a right angle."

Each response assigns to a level. The student at Level 1 thinking gives answers based on a visual model and is identifying the rectangle by its overall appearance. At Level 2 thinking, the student is aware that the rectangle has properties; however, redundancies (i.e., properties that can be derived from other properties) are not noticed. A student's thinking operating at Level 3 attempts to give a minimum number of properties (i.e., a definition), and finally, at Level 4 thinking, a student seeks to prove the fact deductively.

Micheal de Villiers (1987) identifies "six geometric thought categories" including recognition and represent of figure-types (Level 1), use and understanding of terminology (Level 2), verbal description of properties of figure-types (Level 2), hierarchical classification (Level 3), one step deduction (Level 3) and longer deduction (Level 4). Taking an intermediate position, Gutierrez and Jaime (1998) identify each process (recognition, definition, classification and proof) as a component of two or more van Hiele levels of reasoning. They argue that how a student considers and uses the processes is an indicator of the student's level of reasoning at each van Hiele level. For instance, at Level 1, recognition is limited to physical, global attributes of the figures. Students sometimes use geometric vocabulary, but these terms have a visual meaning more than a mathematical meaning. When students describe a rectangle, some of them use the term "perpendicular" for a side when they mean "vertical." In other cases, they are able to notice some mathematical properties of figures, but they are simple properties, such as the number of sides. However, Gutierrez and Jaime also note that the ability of recognition does not discriminate among students in the van Hiele levels 2, 3, or 4.

More recently, Battista (2009) elaborates and refines the van Hiele levels with regard to students' geometric reasoning. The descriptors of the levels he suggests are visual-holistic reasoning, descriptive-analytic reasoning, relational-inferential reasoning, formal deductive proof, and rigor (pp. 92-94), referring to Levels 1 through 5, respectively. For instance, at Level 1 (visual-holistic reasoning), students argue that a square is not a rectangle because a rectangle is "long," or claim that two figures have the "same shape" because they "look the same" (p. 92). At this level, students' justifications of an argument are vague and holistic. At Level 2 (descriptive-analytic reasoning), students would assert that a square is a rectangle because "it has opposite sides equal and four right angles." At this level, students are able to explicitly specify shapes by their parts and spatial relationships among the parts; however, they describe parts and properties informally and imprecisely using strictly informal language learned from everyday life. At Level 3 (relational-inferential reasoning), students start with empirical inference to reason that if a quadrilateral has four right angles (and this is a rectangle), its opposite sides have to be equal because by drawing a rectangle with a sequence of perpendiculars, they cannot make the opposite sides unequal. They then use logical inference to recognize the classifications of shapes into a logical hierarchy (p.94).

These descriptors not only provide detailed information about how researchers identify students' levels of geometric thinking, but more importantly shed a light on ways thinking has been communicated through reasoning and the language skills that students need to develop at each van Hiele level. When conducting studies using the van Hiele model, some researchers use clinical interviews, while others prefer open-ended survey tests. Among all the van Hiele studies, Usiskin's Cognitive Development and Achievement in Secondary School Geometry (CDASSG) project (Usiskin 1982; Senk 1983), and Burger and Shaughnessy's "Oregon Project" (1986) are two of the most frequently used and cited. In the following section, the method of inquiries used in the van Hiele studies are summarized, and some important findings are provided beginning with these two projects.

The Usiskin CDASSG project (Usiskin, 1982; Senk, 1983) used a pre-test and post-test involving four tests to assess 2699 students in full year geometry classes from 13 public high schools in five states. The four tests included Entering Geometry Test, van Hiele Geometry Test, Comprehensive Assessment Program Geometry Test, and Proof Test. The pre-test including Van Hiele Geometry Test and Entering Geometry Test was conducted during the first week of school, and the post-test including van Hiele Geometry Test, Proof Test, and Comprehensive

Assessment Program Geometry Test was conducted three to five weeks before the end of the school year.

The van Hiele Geometry Test was designed to predict students' van Hiele levels at the beginning and the end of the school year. This test consists of 25 multiple-choice items, with five foils per item and five items per level, and was designed to capture the key characteristic of each van Hiele level. In order to develop a rigorous test instrument that describes van Hiele levels in sufficient detail, researchers in the CDASSG project first reviewed nine original works of the van Hieles, including four originally written in English and five translated into English from Dutch, German or French. They compiled all the quotes from the van Hieles' writings that describe behaviors of students at a given level. As an example, the following is a selected list of Level 1 behaviors quoted from the van Hiele writings that Usiskin (1982) provided in the CDASSG project report:

Level 1 (also known as base level, level 0)
1. "Figures are judged according to their appearance."
2. "A child recognizes a rectangle by its form, shape"
3. "The rectangle seems different to him from a square."
4. "A child does not recognize a parallelogram in a rhombus."
5. "A student was able to produce these figures without error…"

The van Hiele Geometry Test instruments were based on the descriptions of students' behaviors at each given level. For example, Items 1-3 were derived from the description "figures are judged according to their appearances"; Item 4 was derived from the description "the rhombus is not a parallelogram. The rhombus appears as something quite different." Figure 2.2 presents one van Hiele Level 1 item. In Usiskin's study, students' responses to the van Hiele Geometry Test were graded using 3 out of 5 corrections and 4 out of 5 corrections. Usiskin and his colleagues compared the two grading methods criterions using the analyses of Type I and Type II error. The statistical analysis showed that depending on whether one wishes to reduce Type I or Type II error, the 3 of 5 correction method minimizes the chance of missing a student and yields an optimistic picture of students' van Hiele levels. The

4 of 5 corrections method minimize the chance of a student being at a level by guessing (see Usiskin, 1982).

Question 4: Which of these are squares?

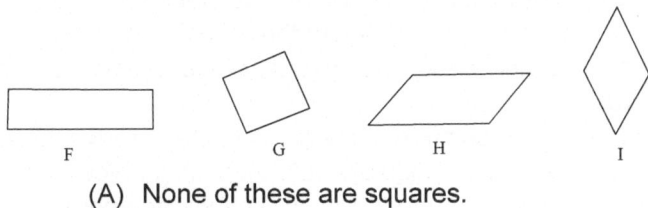

F   G   H   I

      (A) None of these are squares.
      (B) G only
      (C) F and G only
      (D) G and I only
      (E) All are squares

Figure 2.2. An example of a Level 1 test item.

Based on students' test responses, the students were assigned a weighted sum score according to the following:
    1 point for meeting criterion on items 1-5 (Level 1)
    2 points for meeting criterion on items 6-10 (Level 2)
    4 points for meeting criterion on items 11-15 (Level 3)
    8 points for meeting criterion on items 16-20 (Level 4)
    16 points for meeting criterion on items 21-26 (Level 5)

The points were added to give the weighted sum, and the weighted sums were calculated to allow a person to determine upon which levels the criterion had been reached from the weighted sum alone. For example, a score of 19 points indicates that the student had reached criterions on Level 1 (1 point), Level 2 (2 points) and Level 5 (16 points). The assigning of levels, however, was as follows: If a student met the criterion for passing each level up to and including level n and failed to meet the criterion for all levels above, then the student was assigned to level n; if the student could not be assigned to any level, then that student was not said to fit. Thus a student with a weighted sum of

1+2+16 =19 would satisfy the criterion at Level 1, Level 2, and Level 5 and was assigned to van Hiele Level 2 (p. 25). The CDASSG project used Hoffer's (1981) descriptors, labeling the levels as recognition, analysis, ordering, deduction and rigor, from Levels 1 to 5. Additionally, the project reported results using both the classical theory (i.e., all five van Hiele levels are considered) and the modified theory (i.e., Level 5 is excluded from consideration) to classify students into van Hiele levels (Usiskin, 1982).

This large-scale research study showed that nearly 40% of students in the United States finish high school functioning below van Hiele Level 2 (Analysis). Students entering high school geometry courses with higher van Hiele levels, such as Level 2 or Level 3 (Ordering), were more likely to succeed in writing proofs by the end of the school year (Senk, 1983; 1989). Of those studied, students who entered geometry courses functioning at van Hiele Level 1 had a 30% chance of success in proof writing. Entering geometry at Level 2 provided students with a 56% chance of success at proof writing, and all students entering at Level 3 experienced success at proof writing by the end of the school year. These results show that high school students' achievements in writing proofs are positively related to van Hiele levels of geometric thinking and to achievement on standard non-proof geometry content (p.318). The study also concluded, "In the form given by the van Hieles, Level 5 either does not exist or is not testable. All other levels are testable" (Usiskin, 1982, p.79).

These data provide valuable information to determine the initial status of students' geometric backgrounds and to assess their learning progress. However, there have been questions and doubts about the feasibility of measuring reasoning by means of items, and about the internal consistency of the items (Crawley, 1990; Wilson, 1990). In addition, one might question what information might be missed in a paper-pencil test, and how the details of students' thinking processes might be better detected. Nevertheless, the main advantage of this method is that it can be administered to many individuals, and it is easy and quick to distinguish between the thought levels of students.

In contrast, Burger and Shaughnessy's Oregon project (Burger & Shaughnessy, 1986) used clinical interviews to determine students' van Hiele levels. They interviewed 45 students from five school districts in three states, ranging from elementary to middle to high schools. The interviews consisted of eight tasks focusing on geometric shapes, and those tasks were designed to reflect the descriptions of the van Hiele levels.

The design of the interview tasks included drawing shapes, identifying and defining shapes, sorting shapes (e.g., triangles and quadrilaterals); the interview protocols were designed to engage participants in both informal and formal reasoning about geometric shapes. Six of the eight tasks, focused on drawing, identifying, and sorting geometric figures, were expected to elicit the characterizations of van Hiele Levels 0-2 from the protocols. To give an example of the design, Figure 2.3 shows two tasks that were used, *Identifying and defining* (2.3a) and *Sorting* (2.3b).

2.3a. Quadrilaterals   2.3b. Triangles

Figure 2.3. Two experimental tasks from the Oregon Project.

Identifying and Defining
Students were given a sheet of quadrilaterals (see Figure 2.3a), and they were asked to write an S on each square and an R on each rectangle, and if the student was familiar with the terms, a P on each parallelogram and a B on each rhombus. During the interviews, students were asked to justify their reasoning. In the defining part of the interview, the student was asked, "What would you tell someone to look for in

order to pick out all the rectangles on a sheet of figures?" Or, an equivalent question was asked, "Could you make a shorter list? Is this a rectangle? Is this a parallelogram?"(p. 34).

Sorting
A set of cut out triangles was spread out on the table (see Figure 2.3b). The student was asked, "Can you put some of these together that are alike in some way? How are they alike? Can you put some together that are alike in a different way? How are they alike?" (p. 34). This line of questioning was continued as long as the student could come up with new sorting strategies. These activities sought to explore the student's definitions and class inclusions.

The project collected and analyzed students' original written works during the interviews, and the dialogs between interviewers and the students were analyzed and documented as well. For example, on the Drawing Triangles task, interviewees were asked to draw "different" triangles. Based on the interviewees' drawings during the interviews, Burger and Shaughnessy found that for Bud, a 5th grade student, "different triangles" meant triangles in different orientations or positions only. In contrast, for Amy, an 8th grade student, "different triangles" meant having different angle measures and sizes, and for Don, a 10th grade student, "different triangles" meant different types of triangles. Figure 2.4 shows the drawings from Bud, Amy and Don.

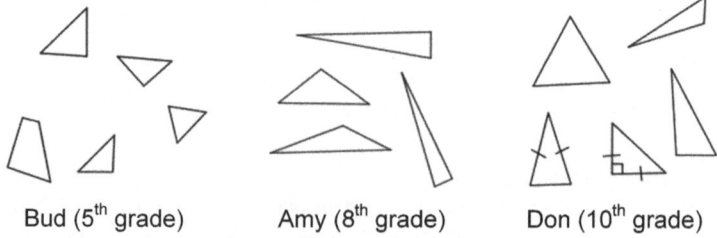

Bud (5th grade)   Amy (8th grade)   Don (10th grade)

Figure 2.4. Bud, Amy and Don's drawings of different triangles.

The "level indicators" developed by Burger and Shaughnessy (1986) describe Levels 0 through 4, respectively, as visualization, analysis, informal deduction, formal deduction, and rigor. Pursuant to students' responses during the interviews, it turned out that, even though all three students were to reason about what is meant by "different triangles," and could provide drawings, all three students were later assigned to three different van Hiele levels: Bud (Level 0), Amy (Level 1) and Don (Level 2). This example illustrates use of the "same language" but very different reasoning. Burger and Shaughnessy also documented the original scripts in which questions were asked for interviewees to complete the task. For example, for the activity "Drawing Triangles," interviewees were asked to draw a triangle (called No.1), and then another triangle (called No.2) that is different in some way from the first one. After the interviewee had done so, he/she was asked to draw a third triangle that was different from the first two triangles, and so on. Later, the interviewees were asked questions such as "How is #2 different from #1?" and "How would they be all different from each other?" (p. 37).

Burger and Shaughnessy's (1986) project confirmed the hierarchical nature of the levels. They also found that age is not significantly related to the levels. However, the reviewers of the project had disagreements and experienced some difficulties in assigning a level to students who appeared to be in the transition between Levels 0 and 1 (p. 42).

The interviewees' written works, on one hand, and their verbal responses to the questions on the other hand, combined to increase the reliability of the data and provide strong evidence for how the data were analyzed and interpreted by researchers. The great advantage of clinical interviews is that the information obtained from the interviews results in a deeper knowledge of the ways students reason. However, this study is clinical with a small sample of students representing a very broad range of ages (Kindergarten to College).

The van Hiele Geometry Test, used to distinguish students' van Hiele levels, is effective in getting initial information about students' levels of thinking, and the CDSSAG project showed that it is a well-tested and designed test instrument. The Oregon project, on the other hand, gives an example of how clinical interviews could well detect students' thinking when engaged in informal and formal reasoning about basic geometric shapes.

There are many other studies using the van Hiele theory that pertain to the learning and teaching of geometry. One such study by Fuys et al. (1988) focused on clinical interviews with sequences of instructions known as "Instructional Module Activities" (p. 11). In this project, all subjects were interviewed individually in six to eight 45-minutes sessions as they worked with an interviewer on the Instructional Modules. The participants of the study were selected to reflect the diversity of sixth-grade students from New York City public schools. To categorize the subjects' levels of thinking, the interviews focused on their progress (or lack of it) within the levels or to higher levels, and on learning difficulties as well (p. 78). This study was designed to investigate if instructional modules would help subjects move through the levels. Fuys et al. (1988) also documented the dialogues between interviewers and students. For example, in the assessment of students' understanding of the exterior angle of a triangle, subjects were given an open question of finding a possible relationship among the three angles indicated in Figure 2.5.

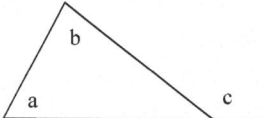

Figure 2.5. An exterior angle of a triangle.

During the interviews, the interviewer gave several prompts to the students such as: "Is any part of angle c related to angle a or angle b?" With the help of the interviewer, the students would sometimes successfully complete the argument of a relationship between two interior angles (angles a and b) and their exterior angle (angle c) of the triangle.

In addition to the clinical interviews, Fuys et al. (1988) also analyzed geometric content of three widely used K-8 textbook series regarding the van Hiele levels. They reported that no more than 2% of the lessons contained content that require geometric thinking at Level 3 (formal deduction), and all of Level 3 lessons appeared at Grades 7 and 8. The remaining 98% represented van Hiele Levels 0, 1, and 2. Analyzing their findings, Fuys et al. concluded that "average students do not need to think above Level 0 (visual) for almost all of their geometry experience through grade 8" (p. 169). In their study of the geometric reasoning of sixth and ninth grade students, Fuy et al. (1988) found the following: 19% of sixth graders performed consistently at Level 0 (visual), 31% performed sometimes at Level 1 (analysis) and sometimes at Level 2 (informal deduction), and the remaining 50% performed sometimes at Level 2 and sometimes at Level 3 (formal deduction). The ninth graders' corresponding percentages were 12% (Level 1), 44% (Level 2), and 44% (Level 3).

Newton (2010) used van Hiele levels to analyze K-8 geometry state standards. Of the 5,710 Grade Level Expectations (GLEs) contained in the K-8 Geometry and Measurement strands of 42 states, 1,667 GLEs (approximately 29%) were labeled as descriptive geometry. The analysis of the descriptive geometry GLEs indicated that approximately 47% of the GLEs are Level 1(visualization), 49% are Level 2 (analysis), and 4% are Level 3 (informal deduction). According to Newton, the absence of Level 3 GLEs in more than 40% of the states and the near absence in the remaining 60% represent the most compelling result of the analyses, since formal deduction (Level 4) is generally expected in high school geometry courses.

The van Hiele model has informed and shaped the improvement of the geometry curriculum (Wirszup, 1976; de Villers, 1999). For example, de Villiers cautioned "no amount of effort and fancy teaching methods at the secondary school will be successful, unless we embark on a major revision of the primary school geometry curriculum along van Hiele lines." In 1999, Clements, Swaminathan, Hannibal and Sarama, encouraged the van Hiele level's use in guiding curriculum development, and suggested "helping children move through these levels may be taken as a critical educational goal" (p. 193). The following year, *Principles and Standards in School Mathematics* (PSSM) cited van Hiele and others who have studied his theory to develop the importance of "… building understanding in geometry across the grades, from informal to more formal thinking" (2000, p. 40). There was an improvement in the Common Core State Stands (CCSSM) as the standards showed more

Level 3 thinking in geometry standards. For example, Dingman, Teuscher, Newton Kasmer (2013) examined CCSSM after it was released, and they found that the proportion of Level 3 standards is higher in CCSSM than the findings from the study of state standards. In CCSSM, the percentage distribution was Level 1 (20%), Level 2 (5%) and Level 3 (30%).

Knowledge of Geometry for Teaching
"Mathematical knowledge for teaching means the mathematical knowledge used to carry out the work of teaching mathematics" (Hill, Rowan & Ball, 2005, p. 373). When suggesting what it means to know mathematics for teaching, Ball, Hill and Bass (2005) argue that teaching involves knowledge of mathematical ideas, mathematics reasoning skills and communication, fluency with examples and terms, and thoughtfulness about the nature of mathematical proficiency. For instance, additional mathematical insight and understanding are required to explain, listen, and examine students' work, and more mathematical analysis is required when correcting students' errors. In addition to mathematical knowledge for teaching, Ball et al. address the need for teachers to have a specialized fluency with mathematical language, and to know what counts as a mathematical explanation.

In this section, research that emphasizes pre-service teachers' knowledge of geometry is reviewed. Mayberry (1983) investigated nineteen undergraduate pre-service teachers' geometric understanding when encountering seven geometry concepts: squares, right triangles, isosceles triangles, circles, parallel lines, similarity, and congruence, all common topics to elementary school textbooks. The study found that two students had difficulty in recognizing a square with a nonstandard orientation (Basic level), while the properties of figures were often not perceived (Level 1). For example, when asked, "Does a right triangle have a longest side?" (p. 60), twelve students responded that they did not think that such a triangle had to have a longest side. With regard to Level 2, the study concluded that class inclusions, relationships, and implications were not perceived by many of the students because they answer the questions for particular figures rather than generalized ones.

Responses to questions about congruence (Level 3) show that fifteen out of the nineteen prospective teachers believed that two right triangles with ten-centimeter hypotenuses are always congruent. Also, ten of nineteen assert that two circles with ten-centimeter chords are always congruent (Table 2.1).

Table 2.1. Responses to Relation Questions about Congruency

Figure: "Are these congruent "	Always	Sometimes	Never	Don't know
A square and a triangle	0	1	17	1
Two squares with 10-cm sides	16	2	0	1
Two right triangles with 10-cm hypotenuses	15	3	0	1
Two circles with10-cm chords	10	8	0	1
Two similar triangles	3	11	3	2

Findings suggested that the pre-service teachers in the study did not yet perceive some of the properties of basic geometric shapes and they did not perceive a proof as a logical chain leading from the "given" to the "conclusion."

Hershkowitz, Bruckheimer and Vinner (1987) conducted a study in Israel of 518 students (grades 5-8), 142 pre-service teachers, and 25 in-service teachers. Findings showed that "teachers have patterns of misconceptions similar to those of students in grades 5-8" (p. 222). More specifically, they assessed 142 pre-service elementary teachers' understanding of geometry concepts in the context of angles, altitudes of triangles, and diagonals of polygons. For example, one of the tasks was to assess the understanding of the angle concept by recognizing the drawing of an angle on a sheet of paper. Results suggested that sixty-eight percent of the pre-service teachers had a proper understanding of the concept of angle (p.223). After assessing pre-service teachers' understanding of the diagonals of polygons,

Hershkowitz, Bruckheimer, and Vinner (1987) found that most of them only drew diagonals that were inside the polygons and rejected the possibility of an exterior diagonal (see Figure 2.6). This result suggests that most pre-service teachers did not use definitions as their primary tool when working with these tasks. They tended to use their own individual image of a given concept (e.g., concept of diagonals of a polygon). These pre-service teachers' individual concept images were misleading in the case of the diagonals of concave polygons.

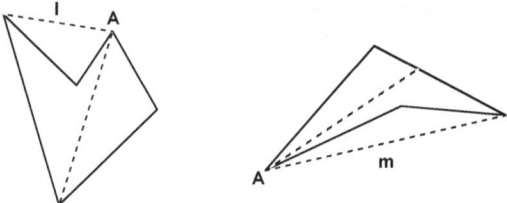

Figure 2.6. The "diagonals" of concave polygons.

Gutierrez, Jaime, and Fornruny (1991) conducted a study to evaluate 32 pre-service teachers' spatial reasoning in three-dimensional geometry. Nine items of the Spatial Geometry Test were grouped into five activities. These activities were designed to elicit pre-service teachers' conceptual understanding of three-dimensional figures, by paying attention either to visual qualities or to the properties of basic geometric shapes such as squares and parallelograms. Activities that asked pre-service teachers to select solids, based on given properties, focused on how they use definitions and properties to identify the polyhedron from a given set of solids. With regard to the question involving writing the differences and similarities between a cube and the Solid I (see Figure 2.7), this activity focused on observation and manipulation of the polyhedron. In response to this question, for instance, one pre-service teacher argued, "In both solids [a cube and Solid I] the faces are parallelograms and both have six faces. And the differences were, the angles in solid I are not right" (p.246). Another pre-service teacher argued, "Solid I and a cube were alike because both solids have parallel faces and all edges are the same, and they differed because they don't have the same shape" (p. 246).

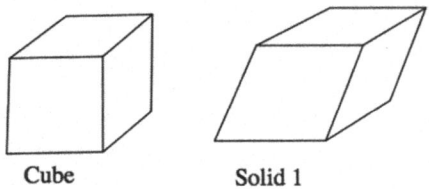

Cube   Solid 1

Figure 2.7. A cube (left) and a solid (right).

These two responses both reasoned about the differences and similarities between a cube and a solid, but the arguments provided by the two pre-service teachers were quite different. The first response focused on the properties of the geometric figures, whereas the latter response mainly paid attention to the visual qualities of figures. In their study, Unal et al. (2009) suggested that pre-service teachers with low spatial reasoning scores might have difficulties developing their geometric knowledge.

Based on Vinner and Hershkowitz's (1987) notion of concept image, Gutierrez and Jaime (1999) conducted a study of pre-service elementary teachers' understanding about the concept of the altitude of a triangle. They analyzed 190 pre-service teachers' written tests, focusing on concept images, difficulties, and errors related to the concept of the altitude of a triangle. Analysis showed that there were more correct responses in the test containing the definition of the altitude than in the test without the definition. This result suggested that the definition seemed to provide information helping these pre-service teachers to improve their understanding of the concept of altitude. Gutierrez and Jaime's also found that there was confusion and misunderstanding between the concepts of altitudes and medians of a triangle, and the concepts of altitudes and perpendicular bisectors of the sides of a triangle. These misunderstandings could be reasons why the pre-service teachers responded incorrectly to a partial image that excludes external altitude (Figure 2.8a), and to a partial image that does not take into account the length of the altitude (Figure 2.8b).

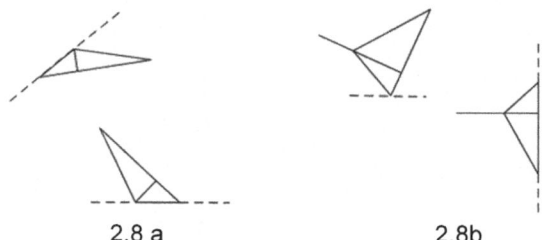

2.8 a   2.8b

Figure 2.8. Responses regarding the altitude of a triangle.

Another issue worth mentioning is the influence of the position of a triangle because of a different rotation of the figure. The easier item appeared to be the prototypical position with a vertical altitude, which suggests that prospective teachers' concept images of altitudes of triangles were limited to only certain types of triangles, and with certain orientations of how triangles are positioned.

More recently, research suggests that pre-service teachers are not adequately prepared for teaching geometry in many different countries (Jones 2000; van der Sandt 2007). In the United States, only 10 percent of elementary mathematics teachers have had courses in each of five areas, including number and operations, algebra, geometry, probability, and statistics recommended by the National Council of Teacher of Mathematics (NCTM). According to a report of the 2012 National Survey of Science and Mathematics Education, the typical elementary teacher has had coursework in only 1 or 2 of these 5 areas, and geometry is one of the areas that is usually missed (Banilower et al, 2013). In the UK, geometry was the topic that pre-service teachers believed they had learned the least (Jones, Mooney and Harries, 2002). In terms of geometric thinking, Samara and Clements (2009) found that the majority of pre-service teachers think at van Hiele Level 1, while Fujita and Jones (2006) reported that UK teachers were at the best of Level 2 thinking. Many pre-service teachers in South Africa were not yet at the Level 3 thinking after months of geometry coursework. In a recent study, Ada and Kurtulus (2010) found that 126 third-year pre-service elementary teachers understood algebraic representations of geometric translation and rotation, but they did not understand the geometry meaning of them.

Li (2013) studied pre-service secondary teachers' understanding of transformational geometry during their coursework, and found that students' had difficulties describing and identifying geometric figures under translations and rotations.

For many years, research on how students learn geometry has been influenced by various theories with different characterizations of cognition and development, resulting in different implications for the learning and teaching of geometry. There is extensive literature and cumulative knowledge generated over time as researchers have been testing the assumptions and applications of the van Hiele model to better understand students' geometric thinking. Many researchers confirmed the usefulness of the model in describing students' geometry thinking. However, a significant part of the discussion relates to critiques of various components of the van Hiele model, including studies that question whether the levels are sequential, linear and discrete (Clements, 2003). There has been further criticism of the assumption that thinking occurs in levels and that students can be identified as being "at a level" (Fuys et al., 1988; Burger & Shaughnessy, 1986) or as a "label" (Gutierrez Jamie (1998, p.45). The same researchers also found the model lacking in depth, and attempted to add more details to the model to assess students' levels of geometric thinking and the development of their thinking (see Hoffer 1981; Michael de Villiers, 1987; Gutierrez and Jaime, 1998; Battista, 2009). These critiques and efforts for more details about the van Hiele model have addressed the needs to revisit it with different theoretical perspectives.

My attempt to explore student geometric thinking connects the van Hiele model with Sfard's communicational approach of mathematical discourse, viewing each van Hiele level of thinking as its own qualitatively different geometric discourse with greater details. The discourse approach views mathematics as a discursive and embodied practice (Gutierrez, Sengupta-Irving & Dieckmann, 2010), and as a cultural activity (Gee, 2011). The mathematics discourse approach also embraces the multi-semiotic nature of mathematical activity that shifts the monolithic view of mathematical thinking (Moschkovich, 2010). Within the general domain of mathematics discourse, I draw on Sfard's

(2008) communicational approach to mathematical learning because it suggests participating in mathematical discourse is engaging in mathematical thinking. This theoretical perspective provides particular robust notion of mathematical discourse that asserts language is a tool whereas discourse is a broader activity in which the tool is used. In particular, Sfard makes the distinction between viewing discourse as one of many factors that shape learning and discourse as the object of the activity, that is, what is learned. This way, mathematical discourse serves as the unit of analysis and as an index of mathematical learning.

The goal of this study is to contribute to this knowledge base by introducing Sfard's communicational of mathematical discourse as it serves a new theoretical perspective, but also an analytical tool with greater details, to explore geometric thinking and its development.

2.2   Theoretical Framework

In this section, I review relevant mathematics discourse approach in learning mathematics and the need for Sfard's communicational approach in my study.

2.2.1 Learning as Communicating

The van Hieles argue that learning a new concept in geometry is a process of learning a new language, because "each level has its own linguistic symbols" (van Hiele, 1959/1985, p.4). They reveal the importance of language use, and emphasize that language is a critical factor in the movement through the levels. However, the word "language" is not clearly defined in the broad use of it (see van Hiele, 1986). For example, it is not clear whether van Hieles refer to "language" as a list of geometric words such as rectangle, square, and parallelogram, or as a set of phrases like "a rectangle is…", "squares could be…" and "parallelograms are…" that describe how individuals view geometric shapes, or if they refer to "language" as all written and verbal expressions that students provide to communicate their thinking about geometric shapes.

In a mathematics discourse approach, Moschkovich (2010) argued that the language of mathematics does not mean a list of vocabulary words or grammar rules, but rather the communicative competence necessary and sufficient for competent participation in mathematical discourse. As Vygotsky points out, the salient connection between the development of a concept and the use of a word:

> A real concept is an image of an objective thing in its complexity. Only when we recognize the thing in all its connections and relations, only when this diversity is synthesized in a word, in an integral image through the multitude of determinations, do we develop a concept (Vygotsky, 1997, Vol. 4, p. 53).

Standing on the shoulder of Vygosky, Sfard (2008) makes a distinction between language and discourse—language is a tool, whereas discourse is an activity in which the tool is used or mediates. This perspective provides an understanding of mathematics as a social and discursive accomplishment in which talk, gesture, diagrams, representations, and objects play an important role. Consequently, mathematics learning requires several modes of communication.

The idea of a discourse approach in the mathematics classrooms is not new. In the past, Ball (1993) delineated a relationship among discourse, content and community to illustrate how these elements helped students to develop what Schoenfeld (1992) called a "mathematics point of view." From this perspective, classroom mathematical discourse is essentially a progress of establishing true claims about mathematics. In her work, Lampert (1998) advanced the case for classroom-based research to consider the impact of language and discourse practices on mathematical learning. For example, Lampert illustrated aspects of "mathematical talk" that included position taking, question asking, proof and justification, expanding ideas, use of evidence, conjectures, symbolic reference, and so on.

Kerslake (1991) focused on language in mathematics classrooms to examine the specific mathematical terms of the content area, and helped to identify how the use of language becomes a resource for understanding students' misconceptions. For example, based on the interviews with students, Kerslake found that students failed to understand fractions as numbers because they perceived fractions as "broken numbers" instead. Moreover, students tended to rely on the everyday language of "sharing" to describe division, and concluded that sharing was likely to have been students' first experience of dividing. Kerslake suggested a closer look at how students think of and talk about fractions in the course of learning.

Moschkovich (2010) also acknowledged the significant role of discourse in learning. Through an analysis of a third grade bilingual mathematics classroom, she extricated two features of mathematical discourse: situated meaning of words (utterances), and focus of attention. She argued that learning mathematics was a discursive activity that involved participating in a community of practice using multiple materials, as well as linguistic and social resources.

Many researchers have attempted to develop frameworks to examine students' discourse in learning mathematics. Sfard's (2008) communicational approach to cognition and to mathematical learning provides a notion of mathematical discourse that distinguishes her framework from others in several ways. She argues that cognitive processes and interpersonal communication are complementary processes in human development. In particular, Sfard (2000) argues that the knowing of mathematics is synonymous with the ability to participate in mathematics discourse. She formulates thinking as an individualized form of communication (not necessarily verbal or visible) and rejects the dichotomies (between thinking and behavior; thought and language; individual and social) inherent in the behaviorist and constructivist approaches. From this perspective, conceptualizing mathematical learning as the development of a discourse and investigating learning means getting to know the ways in which students modify their discursive actions in these three respects: "its vocabulary, the visual means with which the communication is mediated, and the meta-discursive rules that navigate the flow of communication and tacitly tell the participants what kind of discursive moves would count as suitable for this particular discourse, and which would be deemed inappropriate" (Sfard, 2000 p.4).

Sfard's communicational approach implies a participationist stance on the nature of geometric thinking and human interactions in learning mathematics. Therefore, Sfard's discursive framework is grounded in the assumption that thinking is a form of communication and that learning mathematics is learning to modify and extend one's mathematics discourse. The view of mathematical thinking as a form of communication also motivated the van Hieles to develop their model to better communicate geometric thinking with their students. The van Hieles suspected that teachers and their students spoke a very different language because their geometric thinking operated at different levels; therefore, they did not understand each other. The van Hieles developed their model with the insistence on the role of the teacher in changing students' levels of geometric thinking. However, van Hieles' acquisitionist position also implies that thinking is seen as manipulating mental structures, such as conceptions or schemes. Applying Sfard's discursive framework to analyze discourse at each van Hiele level and related to a specific concept would allow me to describe students' geometric thinking from a new perspective. Such a description contributes to the perceived need for details within van Hiele model of geometric thinking.

2.2.2   Sfard's Discursive Framework

Discourse is defined as "any act of communication made distinct by its repertoire of admissible actions and the way these actions are paired with re-actions" (Sfard, 2008, p. 297). A discourse is considered to be mathematical when it features mathematical vocabulary that relates to numbers and shapes. Geometric discourse, a subcategory of mathematical discourse, features mathematical vocabulary specifically relating to geometric shapes, definitions and proofs, etc. (p. 245). The development of discourse involves modifying colloquial mathematical discourse into a more precise mathematical discourse, one that follows meta-discursive rules. For example, geometric shapes are analytically classified by their properties and definitions (meta-level discourse), not just by how they appear holistically (colloquial mathematics discourse). Thus, a stretched out triangle is still a triangle even if it looks distorted. As long as it has three line segments joined at vertices, it is a triangle; and because the segments and vertices can be counted, we engage in a linguistic act (see Sfard, 2008). In Sfard's terms, the mathematical discourse develops from colloquial mathematical discourse; the colloquial mathematical discourse is an important starting point, and to develop mathematical discourse requires a fundamental change in discursive practices.

The mathematical discourses are distinguishable by the four characteristics: *word use*, *visual mediators*, *routines*, and *endorsed narratives*. The four characteristics of mathematical discourse are critical elements to identify if a discourse is "mathematical," and to investigate the development of the discourse. Following is a brief description of each of these.

Word use in mathematical discourse is usually numbers words and comparison words (e.g., bigger, smaller) that appear in the utterance discussing numbers and shapes. Geometric discourse deals geometric shapes (e.g., triangles and quadrilaterals) and their properties. In this current study, words such as *square*, *parallelogram*, *rectangle*, *angle*, *side*, *parallel*, *diagonal*, and their use are investigated. Sfard argues that word use is an all-important matter because being tantamount to what others call word meaning, it is responsible for how the user sees the world, and it is one of the distinctive characteristics of discourse. A student's word use reflects his/her levels of geometric thinking and concept development (as measured in van Hiele levels).

Visual mediators are objects that are operated upon as a part of the process of communication. In a colloquial discourse, visual mediators are images of material things existing independently of the discourse, whereas in mathematical discourse, visual mediators are often involved with symbolic artifacts, created especially for the sake of this particular form of communication. Communication-related operations on visual mediators often become automated and embodied. For example, "the procedures of scanning the mediator with one's eyes with some experience, this procedure would be remembered, activated, and implemented in the direct response to certain discursive prompts, as opposed to implementation that requires deliberate decisions and the explicit recall of a verbal prescription for these operations" (Sfard, 2008, p. 134). In geometry discourse, visual mediators are considered as lines, angles, an orientation of a figure, the shape and the size of a figure, a right angle sign, etc.

Narrative is defined as "a set of utterances, spoken or written, that is framed as a description of objects, of relations between objects or processes with or by objects, and which is subject to *endorsement* or rejection, that is, to being labeled as *true* or *false*" (Sfard 2008: p.300). *Endorsed narratives* are sets of propositions that are accepted and labeled as *true* by the given community. In geometric discourse, endorsed narratives are known as mathematical theories, including definitions, proofs, axioms, and theorems. The statement "a parallelogram is a quadrilateral with two pairs of parallel sides" is an endorsed narrative of parallelogram, defining what a parallelogram is mathematically. Mathematical discourse is conceived as one that should be impervious to any considerations other than purely deductive relations between narratives. In this study, the narratives are those utterances produced by pre-service teachers when identifying and classifying basic geometric shapes, whereas the endorsed narratives are the definitions of different quadrilaterals from textbooks that pre-service teachers encounter and narratives they construct or endorse.

Routines are repetitive patterns as a characteristic of the given discourse. Specifically, mathematical regularities can be observed whether one is watching the use of mathematical words and mediators, or follows the process of creating and substantiating narratives about numbers or geometrical shapes. In fact, such repetitive patterns can be seen in almost any aspect of mathematical discourses: in mathematical forms of categorizing, in mathematical modes of attending to the environment, in the ways of viewing situations as "the same" or different, which is crucial for the students' ability to apply mathematical discourse whenever appropriate.

When someone is doing mathematics, or to be more specific, is engaging in a mathematical task in geometry, patterns such as how one is carefully using mathematical words, or how one is following certain steps when substantiating narratives about geometrical shapes, can be observed. In this study, when participants identify and classify polygons, mathematical regularities are noticed through the interactions. It sheds a light on participants' use of mathematical words and it helps to determine if they follow the process of identifying, defining, and substantiating narratives about geometrical shapes.

A *Routine* can be divided into two subsets—the *how* of a routine, and the *when* of a routine (Sfard, 2008, p. 208): The *how* of a routine: a set of meta-rules that determine the course of the patterned discursive performance (the course of routine, or procedure); the *when* of a routine: a collection of meta-rules that determine those situations in which the participants would deem this performance as appropriate. In this study, *the how of a routine* is observed (through interviews), whereas *the when of a routine* is not discovered as it requires observations over a period of time (consistent observations over weeks, months or even years).

2.2.3 The Levels of Geometric Discourse

My approach of investigating student geometric thinking is to analyze each van Hiele level of geometric thinking as a geometric discourse with word use, visual mediators, narratives, and routines, the four characteristics of discourse described earlier. In this approach, I focus on my analysis on how geometric thinking is communicated through students' use of mathematical words (word use), their repetitive patterns of solving geometric problems (routines), the way they describe geometric shapes (narratives) and symbolic objects they use that serve as communicational tools (visual mediators), in discourse specific context. I expect that when a student's geometric thinking moves towards a higher van Hiele level, his/her geometric discourse also develops to a higher level. As discussed in van Hiele levels, each subsequent level of geometric discourse is a meta-discourse of the former one; each level is a product of reflection on the geometric discourse as it has been practiced so far. To investigate the usefulness of this approach, a model based on the theoretical understandings that describes each van Hiele level as a geometric discourse with respect to *word use*, *routines*, *visual mediators* and *endorsed narratives*, is introduced in the following section.

Based on the theoretical understandings, I first translate the van Hiele levels to discursive terms using the quotes from the van Hieles' writings, the same quotes used to design the van Hiele Geometry Test in the CDASSG project (See Appendix A). I take the van Hieles' descriptions of students' behaviors at each level, and then align them into the four characteristics described in the Sfard's framework; such alignment illustrates student geometric discourse at each van Hiele level. During this process of translating, the van Hieles' quotes at each level were reviewed and analyzed into possible characteristics of a mathematical discourse. For example, the quote, "A child recognizes a rectangle by its form, shape," provides information about how a child identifies a figure, what she/he calls a "rectangle," based on its physical appearance. When this quote is translated into discursive terms, the word, "rectangle" signifies a geometric shape (i.e., a shape that we call a "rectangle"), thus, the *word use* here is a name or a label of the figure. The phrase, "recognizes… by its form, shape" suggests that the direct recognition triggers the decision-making, and therefore the *routine procedure* is a perceptual experience and is self-evident (e.g., [it is] a rectangle [because I see it] by its form, shape). *Narratives* are utterances, verbal or written, that describe objects, and/or relations between objects. The narrative statement is "what is said" about the object during the interview or observation; a student with behavior described in the quote is very likely to say, "it is a rectangle because it looks like one." The *visual mediator* in this situation could include a drawing or picture of a four-sided figure looking like a rectangle. The four characteristics (i.e., word use, routine, endorsed narrative, and visual mediators) of geometric discourse at each van Hiele level are introduced in Tables 2.2-2.3.

Table 2.2. Discursive Translation of van Hiele Levels 1-3

Key terms	Characteristics of Geometric Discourse
van Hieles' description of Level 1. Figures are judged by their appearance.	
Word use	Naming a figure is matching the figure with its name.
Routines	Direct recognition as a perceptual experience that is self-evident.
Endorsed narratives	Descriptions of how one perceives. "This one (square) looks different from this one (a rectangle)."
Visual mediators	2-D geometric shapes, the openness of angles, positions of the lines or physical orientations of a figure are parts of the process of direct recognition.
van Hieles' description of Level 2. Figures are bearers of their properties.	
Word use	Naming a figure is associated with its properties.
Routines	Direct recognition. Object level routines include checking, measuring and comparing partial properties of figures.
Endorsed narratives	Descriptions of visual properties. "Squares are not rectangles because they have all sides equal, but rectangles are not."
Visual mediators	2-D geometric shapes, the openness of angles, positions of segments or physical orientations of a figure are parts of the process of direct recognition and identification of visual properties.
van Hieles' description of Level 3. Properties are ordered and are deduced one from another.	
Word use	Naming a figure signifies the realization of the figure regarding its endorsed narratives.
Routines	Including routines at Level 2. Object level routines producing endorsed narratives.

Endorsed narratives	Descriptions of a definition of a figure and actions on a figure. "Rectangle is a parallelogram having four right angles."
Visual mediators	Figures, lines and angles are parts of the process of identifying necessary and sufficient conditions of a definition.

As shown in Table 2.2, the four characteristics in geometric discourse play different roles at each van Hiele level. For example, for *word use* at Level 1, when a student uses the word "parallelogram," it is an act of "labeling" a figure with the name "parallelogram" with direct recognition and it is self-evident as *routines*. However, at Level 2, the word is used to associate the figure with its properties such as angles, parallel lines, equal sides, etc., with direct recognition, but is not self-evident. In particular, *routines* include measuring, checking and comparing properties as a course of actions. It is at Level 3 that the use of the word "parallelogram" signifies a collection of figures that share a common descriptive narrative (i.e., definition of parallelogram).

Table 2.3. Discursive Translation of van Hiele Levels 4-5

Key terms	Characteristics of Geometric Discourse
van Hieles' description of Level 4. Thinking is concerned with the meaning of deduction.	
Word use	Naming a figure signifies the realization of the figure regarding its endorsed narratives and its connections with other figures.
Routines	Using abstract symbols. Abstract level of routines producing endorsed narratives and making connections among them.
Endorsed narratives	Descriptions of abstract relations. Constructions of narratives using deductive reasoning.

Visual mediators	All level 3 visual mediators, plus abstract symbols, mathematical diagrams.

van Hieles' description of Level 5. Figures are bearers of their properties.	
Word use	Naming a figure signifies the realization with its endorsed narratives and connections with other figures in both Euclidean and non-Euclidean geometry.
Routines	Routines are connected with creativity.
Endorsed narratives	Descriptions of abstract relations in both Euclidean and non-Euclidean geometry.
Visual mediators	Mathematical symbols and artifacts used in both Euclidean and non-Euclidean geometry.

This translation of van Hiele levels into discursive terms provides a glance at additional information about what a student might say (*word use* and *narratives*) and do (*routines*), as well as what symbolic objects (visual mediators) are operated upon as a part of the process of communicating their thinking through a discursive approach. Moreover, the translation provides more details on the development of the levels through the development of geometric discourse.

This study connects Sfard's discourse approach to the van Hiele model, and produces a new framework, *the Development of Geometric Discourse* (is discussed in Chapter 4). This framework expands the five van Hiele levels with characteristics of geometric discourse at each van Hiele level, to using discourse to capture the subtle differences between two van Hiele levels as well as the differences within a van Hiele level.

The remaining part of the study is to uses the empirical data to explore the following questions:

1. At what van Hiele levels do pre-service primary teachers operate?
2. What are the characteristics of pre-service primary teachers' geometric discourse at each van Hiele level?

3. What additional (if any) information does Sfard's discourse approach provide with regarding to pre-service primary teachers' levels of geometric thinking?

In the next section, the study's methodology is discussed.

3 Chapter Three: Methodology

The main goal of this study is to use Sfard's discourse approach as an analytic tool to describe students' geometric thinking. Three primary data sources are used: (1) written responses to the van Hiele Geometry Test (used both as a pre-test and a post-test), (2) transcripts and interview video recordings (from two in-depth interviews, the first interview conducted right after the pre-test, and the second right after the post-test), and (3) participants' written statements, answer sheets to the three tasks during the interviews. These data sources are described in greater detail in this section and are followed by the methods used to analyze these data sources.

3.1 Participants

All participants (n=74) in the study were primary pre-service teachers at a mid-western university in the United States. They were enrolled in a measurement and geometry course designed for future primary teacher candidates. Most of them were second and third year college students and a few were fourth year college students. All seventy-four pre-service teachers took the pre-test, and sixty-three of them took the post-test. With the permission of the course instructors, the pre-and post-tests were administered during their class times by the researcher. Twenty-one of the seventy-four pre-service teachers voluntarily participated in the interviews soon after the pre-test, and twenty of those twenty-one pre-service teachers participated in the second part of the interview soon after the post-test.

In the United States, the geometry and measurement course for pre-service primary teachers is designed to introduce mathematical topics in geometry and measurement that are taught in grades ranging from Kindergarten to eighth grade. It was expected that most participants had studied geometry in K-12, and were therefore familiar with the geometric content topics of this study related to triangles, quadrilaterals and informal proofs in geometry. During the geometry and measurement coursework, the pre-service teachers were introduced to triangles and parallelograms. The discussion included the introduction of angles, perpendicular and parallel lines, and the classification of quadrilaterals. In the classification of quadrilaterals, the pre-service teachers studied parallelograms, rectangles, rhombuses, squares, trapezoids, and kites. They were taught to derive new geometric facts from previously known facts using logical arguments. For instance, during a class, the pre-service teachers worked problems and discovered that an unknown angle in a quadrilateral leads to an unknown angle proof; they also learned from a natural computation to deduce a general fact about a quadrilateral. Towards the end of course, the pre-service teachers were introduced to construct proofs for congruent triangles, and to use congruent triangles to verify properties of quadrilaterals. However, it is important to note that what and how these pre-service teachers learned during their geometry and measurement course work are not the focus of this study.

3.2   The van Hiele Geometry Test

As discussed in the previous chapter, many mathematics educators have used the van Hiele model to determine students' levels of geometrical thinking. The van Hiele Geometry Test used in the Cognitive Development and Achievement in Secondary School Geometry (CDASSG) project was chosen because it was carefully designed and tested by the researchers of the project (Usiskin, 1982, p. 19). The test has been used in well over 40 doctoral dissertations and many thesis studies worldwide. The van Hiele geometry test is used in the current study as the instrument for pre-test and post-test to determine the van Hiele levels of the pre-service teachers' geometric thinking.

The van Hiele Geometry Test contains 25 multiple-choice items, distributed into five van Hiele levels: Items 1-5 (Level 1), Items 6-10 (Level 2), Items 11-15 (Level 3), Items 16-20 (Level 4), and Items 21-25 (Level 5). These items are designed to identify students' geometric thinking at five van Hiele levels. For example, Items 1 to 5 of are designed to identify students' thinking related to van Hiele Level 1, at which figures are judged according to their appearance. Items 6 to10 identify students' behaviors related to van Hiele Level 2, at which figures are the supports of their properties. The van Hiele Geometry Tests are given to provide initial information on pre-service teachers' levels of geometric thinking at the two end points: beginning (pre-test) and the end of the semester (post-test). The analyses of the pre-test and post-test help to determine their van Hiele levels and changes in their geometric thinking over a period of time.

3.3 Interview Tasks

The pre-and post-interviews were designed to explore pre-service teachers' geometric thinking through one-on-one interactions. The goals of the interviews were 1) to investigate ways in which pre-service teachers explain and reason in the context of triangles and quadrilaterals and their properties (e.g., angles, sides, etc.), (2) to explore pre-service teachers' ways of verifying their claims and constructing mathematical proofs, and (3) to probe further into their geometric discourses as observed through these mathematical activities.

Two tasks (See Appendix B) were designed for the interviews and they were printed individually on a standard 8 1/2"× 11" sheet of white paper. The first task included eighteen geometric shapes, labeled with alphabet capital letters shown in Figure 3.1. Among these eighteen polygons, thirteen are quadrilaterals, four are triangles, and one is a hexagon. Sixteen of these polygons were chosen from the van Hiele Geometry Test Items 1-5. Two more polygons were added, consisting of a quadrilateral (Q) and a scalene triangle (S) with no pair of sides equal. Many researchers have used these shapes to categorize students' geometric thinking based on the van Hiele levels (e.g., Mayberry, 1983; Burger & Shaughnessy, 1986; Gutierrez, Jaime, & Fortuny, 1991).

3.3.1   Task 1: Sorting Geometric Figures

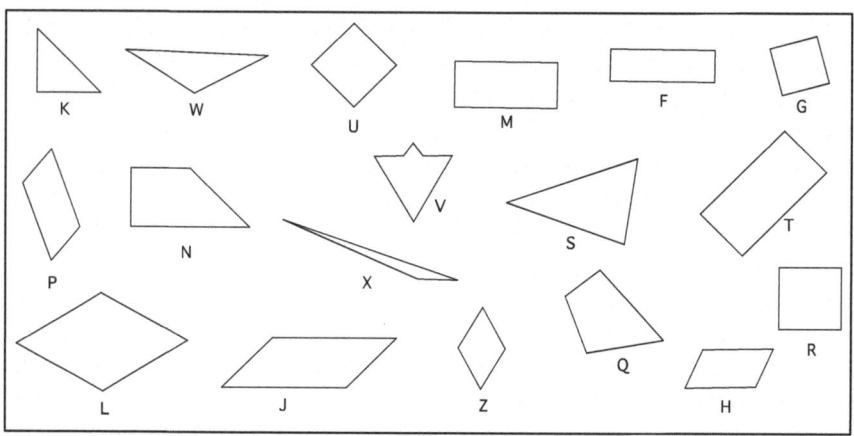

Figure 3.1. Task 1: Sorting geometric figures.

Task 1 was presented to all pre-service teachers at the beginning of the interviews, and each of them was asked to sort the eighteen polygons into groups. After the first round of sorting, each pre-service teacher was asked to regroup and/or subgroup the polygons. For example, some pre-service teachers sorted the polygons into a group of rectangles (see U, M, F, T, R, G in Figure 3.1), and a group of triangles (see X, K, W, S in Figure 3.1). The questions "Can you describe each group to me?" and "Can you find another way to sort these shapes into groups?" allowed pre-service teachers to produce narratives about triangles and quadrilaterals based on their knowledge of the polygons. Analysis of the act of grouping provided information on how participants classify triangles and quadrilaterals.

At the end of Task 1, each pre-service teacher was prompted to write the definitions of rectangle, square, parallelogram, rhombus, trapezoid, and isosceles triangle, and their written narratives were collected. This information revealed how pre-service teachers defined these mathematical terms, and how they made connections between a term and a recognized polygon (e.g., parallelogram), as well as how parallelograms related to one another.

3.3.2   Task 2: Investigating Properties of Parallelograms

Task 2 of the interview had two components. The first component, divided into Part A and Part B, was designed to collect pre-service teachers' drawings of the parallelograms (see Table 3.1), and to gather more information on their knowledge of parallelograms.

Table 3.1. Investigating the Properties of Parallelograms

A. Draw a parallelogram in the space below.
o   What can you say about the angles of this parallelogram?
o   What can you say about the sides of this parallelogram?
o   What can you say about the diagonals of this parallelogram?
B. Draw a new parallelogram that is different from the one you drew previously.
o   What can you say about the angles of this parallelogram?
o   What can you say about the sides of this parallelogram?
o   What can you say about the diagonals of this parallelogram?

Part A began with "Draw a parallelogram in the space below," and then each pre-service teacher was asked to describe the angles, sides and diagonals of the parallelogram. For instance, the question, "What can you say about the angles of this parallelogram?" intended to find out about pre-service teachers' familiarity with the angles of parallelograms. Part B started with "Draw a new parallelogram that is *different* from the one you drew previously," and each pre-service teacher was asked to describe the angles, sides and diagonals of the new parallelogram. This part of the task intended to investigate how pre-service teachers' define parallelograms, and their thinking of *different* parallelograms.

After pre-service teachers completed Parts A and B of the task, they were presented pictures of parallelograms not included in their drawings. Four pictures of parallelograms were prepared for the interviews, each drawn on a 3″× 5″ white index card. Figure 3.2 shows the four parallelograms in the order from left to right consisting of a rectangle, a parallelogram, a square, and a rhombus.

Figure 3.2. Four parallelograms.

These pre-drawn pictures were interview aids to prompt discussion of different parallelograms and their parts. They helped to explore why a pre-service teacher included some parallelograms but excluded others. For example, after a pre-service teacher drew a picture of a parallelogram in Part A and drew a rectangle as a different parallelogram in Part B, she/he was presented pictures of a square and a rhombus, and was prompted to determine if they (square and rhombus) were also parallelograms and why. These additional interview questions were designed to gather more information about pre-service teachers' understanding of parallelograms, and to gain insights on what was missed in identifying and defining quadrilaterals. This task shed light on participants' knowledge of parallelograms, classification of parallelograms, and properties of parallelograms.

Finally, in Task 2, pre-service teachers were engaged in verifying their claims regarding the properties of parallelograms, constructing informal and/or formal proofs. For example, when a pre-service teacher provided a statement that the diagonals of a rectangle were equal, she/he was asked to justify the claim. In order to do so, she/he was engaged in a sequence of conversations with the interviewer by answering questions such as "how do you know?" or "why do you think is true?" and "can you prove it to me?" These questions are designed to learn more about how pre-service teachers' verify mathematical arguments.

3.4   Data Collection

Data collection for this study took place in four phases, as summarized in Figure 3.3.

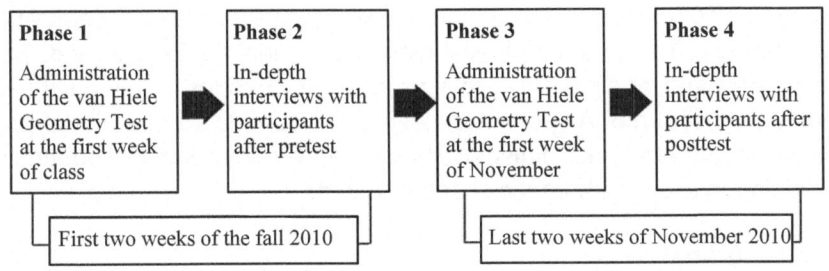

Figure 3.3. Summary of data collection phases.

The first phase of the data collection was the administration of a 35-minutes van Hiele Geometry Test as a pre-test. All pre-service teachers (n=74) enrolled in three sections of the measurement and geometry course took the test during the first class meeting of the fall semester of 2010. All the pre-tests were collected and analyzed, in order to determine pre-service teachers' van Hiele levels of geometric thinking at the beginning of the semester.

In the second phase, twenty-one pre-service teachers voluntarily participated in a 90-minutes in-depth one-on-one interview (pre-interview) with the same researcher a week after the pre-test was given. All pre-interviews were video and audio recorded, and transcribed to serve as the main data recourse for analyzing pre-service teachers' geometric discourses relating to triangles and quadrilaterals. All pre-interviews were completed before the students were introduced to geometric figures in their mathematics content course.

The third phase of the data collection was the post-test, consisting again of the van Hiele Geometry Test. Among the seventy-four pre-service teachers, sixty-three repeated the test ten weeks later during their class meeting. Again, the test responses were collected and analyzed, in order to determine changes in van Hiele levels between the two tests.

The last phase of the data collection consisted of interviews with the pre-service teachers who had participated in the interviews at the beginning of the semester. Among the twenty-one original pre-service teachers, twenty repeated a one-on-one interview (post-interview) with the same researcher and were interviewed for 90 minutes a week after the post-test. The post-interviews were video and audio recorded, and transcribed and analyzed in order to observe changes in interviewees' geometric discourse. All post-interviews were conducted after these pre-service teachers had finished the chapter introducing deductive reasoning in their mathematics content course.

3.5   Data Analysis

The results from the van Hiele geometry test provided general information on pre-service teachers' levels of geometric thinking at the time of the interviews, whereas Sfard's discursive framework served as the analytical framework to explore their thinking in a higher resolution through one-on-one interaction. A brief data analysis is outlined here. First, the pre-service teachers' van Hiele Geometry pre-test and post-test results were analyzed and compared to determine the changes in their geometric thinking according to the van Hiele model as a whole group. When assigning each pre-service teacher to a van Hiele level, the CDASSG project (Usiskin, 1982) test grading method was followed. In particular, one of the CDASSG project's grading method, the 4 of 5 corrections was chosen to determine if a pre-service teacher had reached a given level because it minimized the chance of she/he being at that level by guessing (Usiskin, 1982). When assigning a pre-service teacher to a level, the classical van Hiele levels 0-5 introduced by CDASSG was applied.

The assigning of levels required that the student at level n satisfy the criterion not only at that level but also at all preceding levels. For example, if a participant scored four out of five correct for Levels 1, 2 and 3, two out of five correct for Level 4, and one out of five correct for Level 5, this participant was assigned to van Hiele Level 3 because not only did she satisfy the criterion at Level 3, but at *all* preceding levels as well. However, in this study there were participants assigned as *non-fit* because their van Hiele levels could not be determined from the van Hiele Geometry Test. For example, a participant was assigned *non-fit* because her test results satisfied criterions at the Levels 1, 3 and 5, but not at *all* preceding levels (Usiskin, 1982). All pre-service teachers' overall performances for each item in the pre-test and post-test were examined and analyzed to determine their changes in answering single questions as a group (see the results of both analyses in Chapter 4). The pre-service teachers' van Hiele geometry pre-and post-test results are analyzed and compared to answer the first research question: "At what van Hiele levels do pre-service primary teachers operate?"

The majority of analysis focuses on analyzing pre-service teachers' geometric discourse to support the hypothesis of expanding the van Hiele levels of geometric thinking with Sfard's communicational approach of mathematical discourse. In analyzing pre-service teachers' geometric discourse, the four characteristics of geometric discourse are identified using interview data: (1) Word use, (2) Visual mediators, (3) Endorsed narratives and (4) Mathematical routines. These four characteristics are distinguishable, but they are also interwoven to conceptualize mathematical discourse. For example, mathematical words and their use and visual mediators utilized mathematical objects of mathematical discourse, whereas mathematical routines aimed to produce narratives in given situations. To investigate pre-service teachers' mathematical discourse in the context of triangles and quadrilaterals, the analyses include (1) their word use regarding the names of triangles and quadrilaterals (e.g., rectangle, rhombus, etc.), and the hierarchy of classifications of quadrilaterals, comparing results of the analyses from both interviews, (2) their routine procedures of verifying claims about properties of parallelograms regarding angles, sides and diagonals, comparing results of the analyses from both interviews, and (3) their routine procedures of deriving geometric propositions from other geometric propositions, and in constructing mathematical proofs (see the descriptions of interviews and results of these analyses in Chapter 4). The analyses of the characteristics of pre-service teachers' geometric discourse attempted to answer the second research question: "What are the characteristics of pre-service primary teachers' geometric discourse at each van Hiele level?"

Finally, pre-service teachers' van Hiele levels and their geometric discourse are aligned and compared, to explore their geometric discourse at each van Hiele level (results are provided in Chapter 4). This analysis provides evidence of the usefulness of Sfard's discursive framework as an analytical tool in describing pre-service teachers' geometric thinking at each van Hiele level and the development of thinking through a discursive lens. The analysis also added more information to the model, the *Development of Geometric Discourse*, which serves the purpose of describing pre-service teachers' geometric discourse at each van Hiele level. These analyses are intended to answer the third research question: "What additional (if any) information does Sfard's discourse approach provide with regarding to pre-service teachers' levels of geometric thinking?"

4 Chapter Four: Results

The findings of this study are reported in two sections: the results of the van Hiele Geometry pre-and post-test analyses and the results pre- and post-interview analyses. The results of the van Hiele geometry test analyses provide initial information of sixty-three primary pre-service teachers' van Hiele levels of geometric thinking. The results of individual interview analyses reveal details of pre-service teachers' geometric discourse at each van Hiele level. Both sections are aimed at investigating and describing changes in students' geometric thinking in the context of quadrilaterals.

4.1 Results from the van Hiele Geometry Test

Table 4.1 shows how about forty-eight percent of pre-service teachers reached Level 3 (n=30) at the post-test, and that number almost doubled over the pre-test.

Table 4.1. Distributions of Pre-Service Teachers' van Hiele Levels on Both Tests

Level	VHB		VHE	
	N	%	N	%
0	6	9.52	4	6.35
1	8	12.70	9	14.29
2	9	14.29	8	12.70
3	17	26.99	30	47.62
4	2	3.17	2	3.17
5	0	0	1	1.59
Total fitting	42	66.67	54	85.71
Non-fit	21	33.33	9	14.29
Totals	**63**	**100**	**63**	**100**

Note: VHB indicates van Hiele levels at the beginning of the semester, whereas VHE indicates their van Hiele levels ten weeks later.

Among the sixty-three pre-service teachers, thirty-eight of them were able be assigned to a van Hiele level in both tests according to their van

Hiele geometry test results. Among these thirty-eight pre-service teachers, sixteen of them (42%) moved up at least one van Hiele level. One pre-service teacher moved up to Level 5 at the post-test. There was a slight reduction at Level 0 and Level 2, and a slight increase at Level 1 from the pre-test to the post-test. There were twenty-one pre-service teachers assigned to *non-fit* at the pre-test, but it was reduced to nine at the post-test. The reduction of the *non-fit* test result suggests that the responses to the van Hiele Geometry Test were more consistent at the post-test than the pre-test. The numbers show that a little more than half (52.4%) of pre-service teachers reached Level 3 or above at the post-test.

The van Hiele Geometry Test results show an increase in correct answers for Items 1 to 15 from the pre-test to the post-test, indicating that as a group, these pre-service teachers did better at the post-test in items related to van Hiele Levels 1, 2 and 3. However, there was an increase in correct answers for some items at Level 4 and Level 5, but it is not consistent. Figure 4.1 compares these pre-service teachers' performance based on the number of correct answers for each item at the pre-test and the post-test.

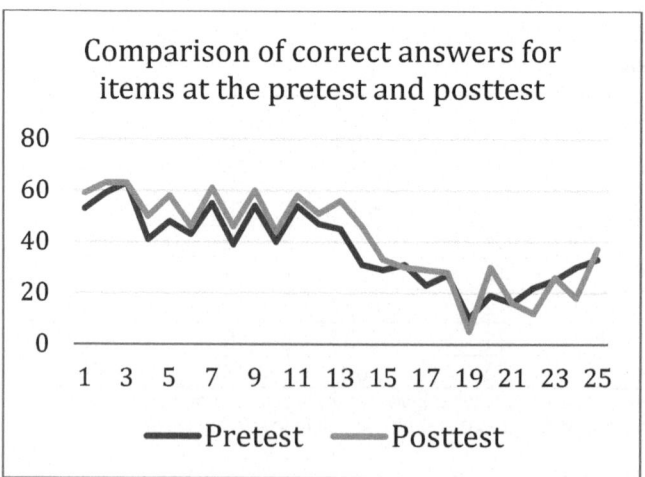

Figure 4.1. Comparison of correct answers for items at both tests.

Overall, these pre-service teachers (n=63) had a better performance on these items at the post-test than at the pre-test. In particular, they were better in answering questions relating to van Hiele levels 1 to 3 at the post-test than at the pre-test. However, there was a reduction in correct answers starting at Item 16, and that continued intermittently to Item 25. Items 16 to 25 are designed to identify pre-service teachers' Level 4 and Level 5 thinking. In looking at these results, it is evident that these pre-service teachers were getting more familiar with triangles and quadrilaterals, as well as with properties related to these polygons. However, it is not clear that they improved in doing proofs and thinking at an abstract level, mathematical activities related to van Hiele Level 4 and Level 5. For example, Item #19 uses the language of logic, with one of the choices stated as, "Every term can be defined and every true statement can be proved true." For most of these students (primary pre-service teachers), they were not familiar with the language in logic, and might not know what the statements of the items #19 meant. Thus, prevent them to get the correct answers.

Among the sixty-three pre-service teachers who completed pre-and post-van Hiele tests, twenty voluntarily participated in the interviews conducted after the pre-test and the post-test. Table 4.2 summarizes distributions of the twenty pre-service teachers' van Hiele levels at the pre-test and the post-test.

Table 4.2. Distributions of Interviewees' van Hiele Levels at Both Tests

Level	VHB		VHE	
	N	%	N	%
0	2	10	1	5
1	1	5	0	0
2	2	10	4	20
3	6	30	10	50
4	1	5	2	10
5	0	0	0	0
Total fitting	12	60	17	85
Not fit	8	40	3	15
Totals	20	100	20	100

The percentages at each van Hiele level of the twenty pre-service teachers matched closely with the corresponding percentages of the whole group (n=63). For example, thirty percent of them were at Level 3 in the pre-test, whereas it increased to fifty percent at the post-test. These numbers are close to those for the whole group—twenty-eight percent and forty-eight percent, respectively (see Table 4.2). Thus, it appears that the twenty pre-service teachers who participated in the interviews are a good sample size for the interview portion of the study.

The van Hiele Geometry Test provided initial information about participants' van Hiele levels at the time of the study, but it did not reveal the dynamic changes in participants' levels of thinking. For a deeper analysis of participants' thinking process, twenty participants were interviewed soon after the pre-test and post-test. The results of the interview analyses are provided in the next section.

4.2   Changes in Geometric Discourse

This section focuses on the analyses of five pre-service teachers' geometric discourses: Molly, Ivy, Kevin, Sam, and Judi (pseudonyms). The van Hiele geometry pre- and post-tests suggested changes in Molly (Level 1-Level 3), Ivy (Level 0-Level 3) and Kevin's (Level 3-Level 4) geometric thinking, and indicated no change in Sam (Level 2-Level 2) and Judi's (Level 3-Level 3) geometric thinking. Kevin was a first-year college student, and the other four pre-service teachers were second-year college students at the time of the study. These pre-service teachers are chosen because they represent a range of van Hiele levels from Level 0 to level 4 and various cases of geometric discourses across different van Hiele levels. Table 4.3 summarizes the five pre-service teachers' van Hiele levels and some of their academic information.

Table 4.3 Summary of Five Pre-Service Teachers' van Hiele Levels

Name	VHB	VHE	Major	Year
Ivy	3	3	Primary Education	Second year
Kevin	3	4	Primary Education	First year
Molly	1	3	Special Education	Second year
Sam	2	2	Primary Education	Second year
Judy	3	3	Early Childhood	Second year

The analyses of pre-service teachers' geometric discourses in the context of quadrilaterals and triangles focus on *word use* and *routines*. *Word use* includes the use of the names of polygons (e.g., rectangles parallelogram, etc.), and the names of the parts of polygons (e.g., angle, side, etc.). The *routines* include *routines* of *sorting, identifying, defining, conjecturing,* and *substantiating* as course of actions while these pre-service teachers engaged in solving geometric tasks during the interviews.

Recall that a *routine* is a set of meta-rules that describes a repetitive discursive action. In this study, different routines such as recalling routines, identifying routines and defining routines are involved given the nature of the tasks. For example, the *routine of sorting* is a set of routine procedures that describes repetitive actions in classifying polygons (e.g., by their family appearances, by visual properties, etc.). The *routine of identifying* is a set of routine procedures that describes repetitive actions in identifying polygons (e.g., by visual recognition, by partial properties check, etc.), whereas the *routine of defining* is a set of repetitive actions related to how polygons are described or defined (e.g., by visual properties, by mathematical definition, etc.). In endorsed narratives such as mathematical definitions or axioms, the *routine of recalling,* a subcategory of the *routine of defining,* is a set of repetitive actions using previously endorsed narratives and "it can indicate a lot not just about how the narratives were memorized, but also about how they were constructed and substantiated originally" (Sfard, 2008, p. 236).

When engaging in mathematical tasks, "guessing and checking" are seen as common activities because students' initial response often triggered by visual recognition. Therefore, the *routine of conjecturing* is a set of repetitive actions that describe a process of how conjecture is formed through "guessing and checking"; and the *routine of substantiating* is a set of patterns describing a process of using endorsed narratives to produce new narratives that are true. For example, a *routine of substantiation* can be seen as using a valid triangle congruence criterion to proof formally or informally, two triangles are congruent.

Among the twenty pre-service teachers who participated in the interviews, four showed a change in their van Hiele levels from lower to higher according to the van Hiele geometric pre- and post-test results, with Ivy, Molly, and Kevin among the four pre-service teachers. Ivy moved three van Hiele levels from Level 0 to Level 3; Molly moved two van Hiele levels from Level 1 to Level 3; and Kevin moved one van Hiele level from Level 3 to Level 4. The results of Molly, Ivy and Kevin's geometric discourse analyses are provided in the following subsections.

4.2.1 Case One: Changes in Molly's Geometric Discourse

Molly was a second-year college student and took her last geometry class five years prior to the study. The van Hiele Geometry Test showed that she was at Level 1 at the pre-test, and ten weeks later she moved up two van Hiele levels, to Level 3. A summary of Molly's changes in geometric discourse from pre-interview to post-interview is as follows:

- Molly's routines of sorting polygons changed from grouping polygons according to their family appearances to classifying polygons according to their visual properties and definitions.
- Molly's identifying routines changed from visual recognition, which is self-evident, to identifying partial properties of the polygons (i.e. sides and angles).
- Molly's use of the names of parallelograms changed from describing the parallelograms as collections of unstructured quadrilaterals that share some physical appearances, to using the names as collections of quadrilaterals that share common descriptive narratives.

- Molly's substantiation routines were not observed in both interviews. That is, Molly did not use measurement tools to prove or disprove congruent parts of the polygons at the object level nor did she use informal or formal mathematical proofs at the abstract level.

Changes in Molly's Routines
At the pre-interview, Molly stated, "I group them solely on their amount of sides," and sorted the given polygons (n=18) into three groups on her first attempt: 1) 3 sides (K, W, X, and S); 2) 4 sides (U, M, F, G, P, T, L, J, H, R and Z; and 3) Trapezoids (V and Q). See Figure 4.2 for some examples of each group based on Molly's written response.

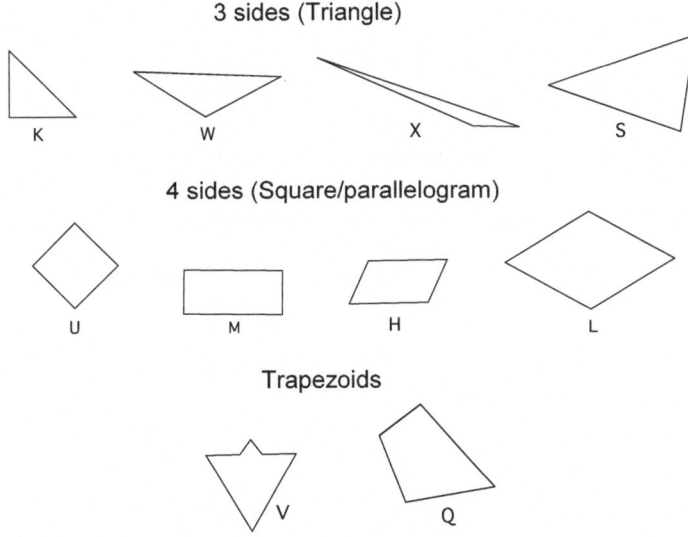

Figure 4.2. Samples of Molly's grouping of polygons in the pre-interview.

Molly included all triangles (n=4) in the 3-sides group and called it the *triangle group.* She included all *squares, rectangles, rhombi* and *parallelograms* in the 4-sides group (n=11), and called it the *square and parallelogram group.* She grouped Q (a quadrilateral) and V (a hexagon) together as a *trapezoid* group because to her, a *trapezoid* was "a figure with five sides, varying in length," and "often make these odd shapes." N (a right trapezoid) was not included in any of these groups shown in Figure 4.2.

The researcher asked Molly to regroup the sixteen polygons differently and Figure 4.3 shows she regrouped triangles according to attributes of angles and sides, into right triangle (K), isosceles triangle (W) and scalene triangles (X and S). Molly also regrouped the 4-sided polygons according to their family appearances, with the names of *squares*, *rectangles*, *parallelograms*, and *rhombuses*. Figure 4.4 shows two of the groups: the *rectangles* and the *squares*.

Triangle:

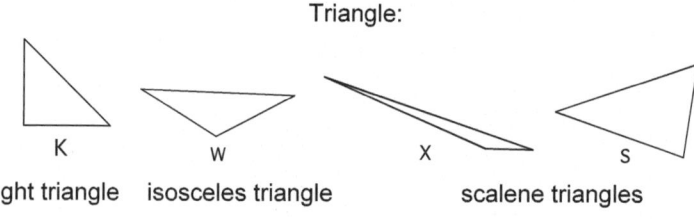

right triangle   isosceles triangle   scalene triangles

Figure 4.3. Molly's regrouping of the triangles in the pre-interview.

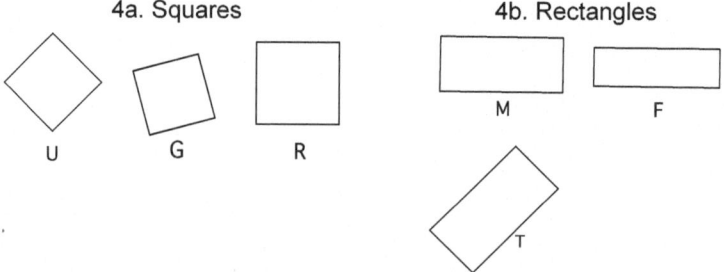

Figure 4.4. Molly's regrouping of the quadrilaterals in the pre-interview.

When asked for an explanation, Molly responded, "I know this figure (U) and this figure (M) are different, but they both belong to the same quadrilateral group." Molly did not include a right trapezoid (N) in any of these groups on her second attempt. To investigate further, the researcher asked Molly to explain if a parallelogram (J) and a right trapezoid (N) could group together, and the following conversation took place:

Researcher: Can we group these two together?

J   N

Molly: I wouldn't believe so... Just because this [N] shows the angle... it doesn't have the properties of a square or a rectangle, [it] has no congruent part.

Molly's routines of sorting polygons at the pre-interview are illustrated in Figure 4.5

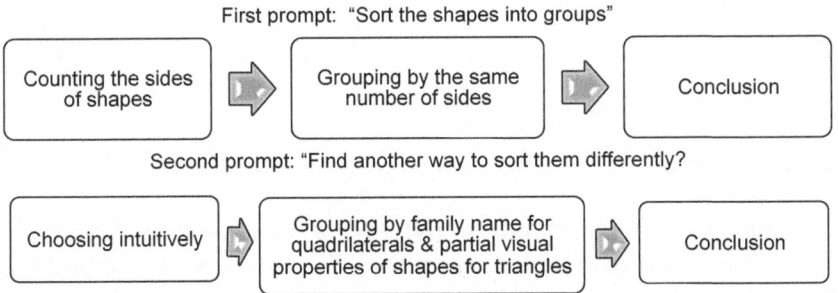

Figure 4.5. Molly's routines of sorting polygons in the pre-interview.

At the post-interview, the same task was performed. Molly grouped the polygons by "looking at the numbers of sides solely," and she sorted eighteen polygons into three groups: 1) *Triangles* (K, W, X and S); 2) *5-sided* (V); and 3) *Quadrilaterals,* all other polygons in the task. Figure 4.6 summarizes Molly's first attempts at both interviews with some key examples.

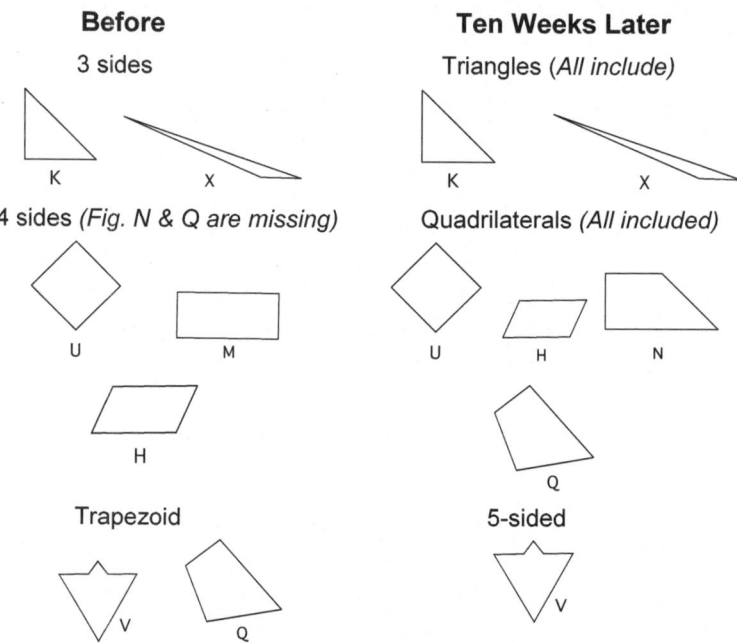

Figure 4.6. A Summary of Molly's grouping of polygons in the interviews.

Molly grouped a right trapezoid (N) and a quadrilateral (Q) in the quadrilateral group using defining routine. For example, she argued, a right trapezoid (N) is a quadrilateral because it is "a four sided figure with one distinct pair of parallel sides." This is a different response than the one she provided at the pre-interview about the right trapezoid (N), "it does have four sides, but... no congruent parts," which was not included in any categories she grouped. Molly's explanations of regrouping the quadrilaterals into squares, rectangles and parallelograms at the post-interview are provided below:

Molly: Quadrilaterals, you know that you have your square [U] because...each forms 90-degree and all the side lengths are equal.

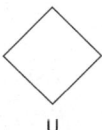

U

Molly: these are rectangles [F and M] because two sides and those two sides are the same. But again they form 90-degree angles

F   M

Molly: opposite angles are equal and opposite sides are equal, so these three [L, J and H] would be an example of parallelogram.

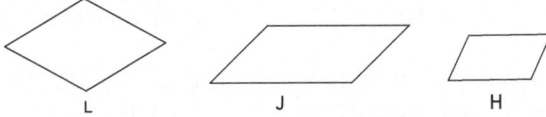

L   J   H

At the post-interview, Molly was able to use her definitions of *square, rectangle, parallelogram,* and *rhombus* to identify and to regroup the quadrilateral group. She regrouped quadrilaterals into: 1) *squares* (U, G, and R); 2) *rectangles* (M, F, and T); 3) *rhombus* (Z); and 4) *parallelogram* (L, J, and H). When the researcher asked her if a square (U) and a right trapezoid (N) could group together, Molly responded:

They can group together as both being same amount of sides...but in terms like property...no...they both have two parallel sides, but a trapezoid cannot be branched off with parallelograms into rectangles and squares....

U   N

This action indicated Molly's ability to compare figures, not only focusing on the "same amount of sides," but also on their visual properties, "they both have parallel sides." Table 4.4 summarizes Molly's routines in response to sorting geometric figures in both interviews.

Table 4.4. A Comparison of Molly's Routines of Sorting Polygons in the Two Interviews

Before	Ten Weeks Later
First prompt: "Sort the shapes into groups"	*First prompt: "Sort the shapes into groups"*
1. Counting the sides of shapes *(Counting)* 2. Grouping by the same number of sides 3. Conclusion	1. Counting the sides of shapes *(Counting)* 2. Grouping by the same number of sides 3. Conclusion
Second prompt: "Find another way to sort them differently"	*Second prompt: "Find another way to sort them differently"*
1. Direct recognition of possible candidates *(Visual recognition)*	1. Direct recognition of possible candidates *(Visual recognition)*
2.a Grouping by family appearance of quadrilaterals and parallelograms *(Visual recognition)*	2.a Grouping by common descriptions of quadrilaterals and parallelograms by visual properties and some mathematical definitions *(Defining routine)*
2.b Grouping by properties of angles and sides in triangles *(Defining routine)*	2.b Grouping by properties of angles and sides in triangles *(Defining routine)*
3. Conclusion	3. Conclusion

Table 4.4 shows no change in Molly's routines of sorting triangles in her first attempt, but there was a change in routines of sorting quadrilaterals on the second attempt. The difference on the second attempt is shaded and it shows that Molly used defining routines to sort polygons at the post-interview. Molly's routine of sorting changed from visual recognition to classifying polygons according to their common descriptive narratives.

Molly's responses to the questions in Task Two also revealed changes in her routines. Task 2 involves two sets of activities about parallelograms, and is designed to investigate pre-service teachers' familiarity with the angles, sides, and diagonals of a parallelogram by asking them to draw two parallelograms that are different from each other (see Appendix B). As shown in Table 4.5, Molly drew a parallelogram and then a rectangle as a different parallelogram in the pre-interview. She declared that the two parallelograms were different because "I would change the sizes of it [side]". Molly described the second drawing as, "it's a rectangle... but it's not the typical looking parallelogram." In response to the questions about the angles of the parallelograms, Molly expressed her frustrations on the angles: "I am still stuck on the question on what it means by the angles, ... Usually when I'm talking about angles, we have measurements...[pausing] I feel like the angles would be the same just based on how it looks."

Table 4.5 Molly's Routines of Verifying on the Angles of Parallelograms in the Pre-Interview

Q: "What can you say about the angles of this parallelogram?"		
	Parallelogram	Rectangle
Conjecture (Guessing)	"I would assume that they are the same for the opposites"	"They would have to be equal...or add up to a certain amount"
Q: "How do you know?"		
Routines	Visual recognition	Visual recognition
Declared Narrative	"No. I don't know." "Just based on how it looks"	"Just looks more like a stereotypical parallelogram"

Molly made intuitive claims about the angles of a parallelogram and a rectangle using direct recognition. For example, Molly assumed that the angles were "the same for the opposites" in a parallelogram using direct recognition. In this case, the question "How do you know [they are the same]?" did not lead to any substantiations of the claim, nor lead her to endorse any narratives using mathematical definitions; instead Molly's conclusion was reached by direct visual recognition, which was self-evident. This routine pattern also appeared when Molly was discussing the diagonals of a parallelogram.

> Researcher: What can you say about the diagonals of this parallelogram?

> Molly: The diagonals would be equal...
> Researcher: How do you know the diagonals are equal?
> Molly: You have to measure and make sure these were, all their sides were the same, right here [the sides], would all equal... on each side all equaling the same parts.

Molly declared a narrative about the diagonals of the parallelogram stating that, "the diagonals would be equal." This is an invalid statement because the diagonals of this parallelogram are not equal, and it can be detected if Molly would use the ruler to check the measurements of the diagonals. However, Molly did not check because her direct recognition was intuitive and also was self-evident. In that case, there was no need to substantiate the narrative, "the diagonals would be equal," but instead Molly made her own intuitive conclusion that the "diagonals were equal" because "...all their sides were ...the same."

Molly's responses to the question "what can you say about the angles of the parallelogram?" in the post-interview are summarized in Table 4.6.

Table 4.6. Molly's Routines of Verifying on the Angles of Parallelograms in the Post-Interview

Q: "What can you say about the angles of this parallelogram?"	
Parallelogram	Rectangle
Declared Narratives "Opposite angles equal and they don't form 90-degreee angle"	"You could say that the opposite angles are equal, and in this one all angles are equal"
Q: "How do you know?"	
Routines a. Visually identify partial properties of a parallelogram by checking the condition of opposite angles *(Identifying routine)* b. Describe a parallelogram with no right angles *(Defining routine-recalling)*	a. Visually identify partial properties of a rectangle by checking the condition of opposite angles *(Identifying routine)* b. Describe a rectangle with right angles *(Defining routine –recalling)*
Declared Narratives "I would just say the property of parallelogram"	"It has properties of parallelogram. It's a rectangle"

Molly did not know how to draw a conclusion about the angles in a parallelogram without measurements at the pre-interview, but she was able to discuss the angles of parallelograms using the properties of a parallelogram (*defining routine*) at the post-interview. For example, when Molly declared the narrative "opposite angles are equal and they don't form a 90-degree angle," she identified that this 4-sided polygon was a parallelogram (*identifying routine*) and described the parallelogram, as it had no right angles using defining routines. Similarly, Molly was able to identify the differences of the angles between two parallelograms: a *parallelogram*, "opposite angles are equal and … they don't form a 90-degreee angle" and a *rectangle*, "the opposite angles are equal, and in this one [rectangle] all angles are equal."

Here, we begin to see the change in Molly's routines of identifying, from visual recognition, to identifying visual properties of the angles in a parallelogram, as well as the use of defining routines to justify claims at the post-interview. At this stage, Molly's routine of defining was more of a recalling. In discursive terms, recalling is a routine that one performs to summon a narrative that was endorsed in the past, or to recall narratives that were memorized in the first place (Sfard, 2008).

During the pre-interview, Molly showed more confidence in discussing the sides of the parallelograms than the angles of the parallelograms. The researcher asked her about the sides of the parallelogram.

> Researcher: What can you say about the sides of this parallelogram?

> Molly: Opposite sides are equal…
> Researcher: How do you know they are equal?
> Molly: Just the properties of a parallelogram. If I measure it out…if I draw it with a ruler, it would have to be the same for each side
> Researcher: Is there a way that you can show me that the opposite sides are equal?
> Molly: I would draw it out with two sides having to be the same measure and these two having to be the same measure. But for one of the opposite sides, they have to be longer than others to not to make it the properties of a square.

In this dialogue, Molly declared a narrative, "opposite sides are equal." When asked for substantiation, Molly justified her claim by saying "just the property of a parallelogram." and provided another explanation, "If I measure it, draw it out with a ruler...it would have to be the same." Molly then verbally described a set of procedures to justify her claim: "draw on with a ruler," and "draw it out with two sides having to be the same measure." Molly's routines of verifying the sides of a parallelogram and a rectangle in the pre-interview are summarized in Table 4.7.

Table 4.7. Molly's Routines of Verifying on the Sides of Parallelograms in the Pre-Interview

Q: "What can you say about the sides of this parallelogram?"	
Parallelogram	Rectangle
Conjecture (Guessing) "Opposites are equal" "Opposite sides … one longer than the other"	"Each opposite side is equal in length"
Q: "How do you know?"	
Routines a. Visual recognition b. Identify partial properties of a parallelogram *(defining routine-recalling)*	a. Visual recognition b. Identify partial properties of a parallelogram *(defining routine-recalling)*
Declared Narrative "Just on the properties of a parallelogram"	"Rectangle can still have the properties of a parallelogram"

Table 4.7 shows Molly identified the equal sides of a parallelogram and a rectangle intuitively and verified her claims using properties of parallelograms. However, Molly's description of a parallelogram, "opposite sides one longer than the other," indicated that her identifying routines were triggered by the visual appearances.

In contrast, the following dialogue during the post-interview indicated some changes in Molly's defining routines as she declared, "Opposite sides are parallel and equal" in referring to the sides of a parallelogram (also a rectangle).

> Researcher: What can you say about the sides of this parallelogram?

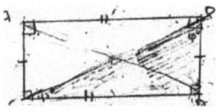

> Molly: Opposite sides are parallel and they should be equal.
> Researcher: Why do you say "should be"?
> Molly: Because it has the properties of a parallelogram. By looking at it, it looks as if they are, so it could be good.
> Researcher: Can you prove that the opposite sides are equal and parallel?
> Molly: Based on the properties of it [parallelogram].

At this point, Molly's course of action consisted of visual recognition "by looking at it" and *visually* checking the sides, to verify that they were parallel and equal. Her reasoning of why opposite sides are equal and parallel was, "the properties of it [parallelogram]." Molly provided a similar response when she discussed the diagonals of the parallelogram.

> Researcher: What can you say about the diagonals of the parallelogram?
> Molly: They bisect whatever angles these are.
> Researcher: How do you know diagonals bisect angles?
> Molly: It's the properties of it [parallelogram].
> Molly's claim about the diagonals of this parallelogram (the one she drew) bisect the angles is invalid, and the only justification she offered was "it's the property of it." The researcher prompted for more reasoning by asking Molly, "Can you prove it [the diagonals bisect the angles]?"
> Molly: The diagonals would have to be equal, form these right triangles, specifically 90-degree. [Draw diagonals]

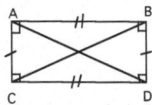

Molly: it takes half of these, so it'd be 45, 45, 90.

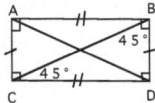

Molly: Two diagonals form two congruent triangles because they have the same base,length, side lengths. [Referring to shaded and un-shaded triangles]

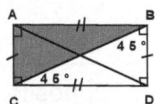

Molly: [Diagonals] bisect the angles.
Researcher: How do you know they [diagonals] bisect angles?
Molly: This is one of our right triangles. [Shade the triangle].

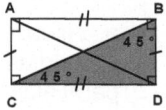

Molly: This is 90-degree angle here. [Pointing at the angle D]

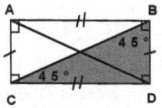

Molly: This diagonal [referring to the hypotenuse of the shaded triangle] completely bisects these two angles (∠B and ∠C) in half because we have our 45-, 45- 90-degree angles. You know that the [angles of] triangle has to be equal to180.

Molly was confident about her conclusions of "diagonals bisect the angles" in a parallelogram (a rectangle). Molly demonstrated incorrect deductive reasoning when she talked about the properties of parallelograms; however, she did use deductive reasoning at the post-interview, not explicitly, to verify the "diagonals form congruent triangles." Table 4.8 summarizes Molly's routine procedures concerning the sides of a rectangle and a square in the post-interview.

Table 4.8. Molly's Routines of Verifying on the Sides of Parallelograms in the Post-Interview

Q: "What can you say about the sides of this parallelogram?"	
Parallelogram (rectangle)	Square
Declared Narratives "Opposite sides are parallel and they should be equal"	"Opposite sides would be congruent" "They are parallel to one another"
Q: "How do you know?"	
Routines a. Visually identify partial properties of a parallelogram by checking the condition of the sides *(Identifying routine)* b. Describe a parallelogram –opposite sides are equal & parallel *(Defining routine-recalling)*	a. Visually identify partial properties of a parallelogram by checking the condition of the sides *(Identifying routine)* b. Describe a particular parallelogram-opposite sides are equal & parallel *(Defining routine-recalling)*
Declared Narratives "It has the properties of a parallelogram"	"It's a basic property of a parallelogram" "To be a parallelogram, opposite sides have to be parallel, making them congruent"

Molly's changes in routine procedures were (1) from visual recognition that was self-evident, to identifying partial properties and using properties about the angles of a parallelogram, (2) from identifying partial properties (i.e., equal sides) in the pre-interview, to identifying more properties (i.e., equal and parallel sides) in the post-interview is summarized in Table 4.9.

Table 4.9. Summary of Molly's Routines in Task Two from Both Interviews

	Pre-Interview		Post-Interview	
	Routines		Routines	
Parts of parallel-grams	Identifying Routine	Defining Routine	Identifying routine	Defining routine
Angles	Visual recognition Self-evident	No	Visual recognition/ Identifying partial property	Recalling
Sides	Visual recognition/ Identifying equal sides	Recalling	Visual recognition/ Identifying equal and parallel sides	Recalling
Diagonals	Visual recognition Self-evident	No	Visual recognition Self-evident	No

Molly's identifying routines changed from direct recognition, to recalling and identifying partial properties of parallelograms when discussing the angles and diagonals of parallelograms. Her substantiation routines were self-evident, and so were neither at an objective level (concrete comparison or measurement) nor at a meta-level. In analyzing Molly's geometric discourse, the analyses also show the changes in her use of geometric words.

Changes in Molly's Word Use

The analyses of Molly's use of mathematical words focus on words related to quadrilaterals, and in particular, parallelograms. One common definition of quadrilateral is "a four sided polygon" (Usiskin, Griffin, Witonsky and Willmore, p.11, 2008). Among all quadrilaterals, six types of quadrilaterals are found predominately in all school geometry texts: *parallelograms, trapezoids, isosceles trapezoids rectangles, squares, and rhombuses*. A common definition of parallelogram is "a quadrilateral with two pairs of parallel sides" (p. 21). By definition, rectangles, squares and rhombuses are parallelograms. Table 4.10 summarizes the frequencies of the names of quadrilaterals Molly mentioned in the interviews.

Table 4.10. Frequencies of Molly's Use of the Names of Quadrilaterals in the Interviews

Name	Frequency				Total Frequency	
	Pre-T1	Post-T1	Pre-T2	Post-T2	Pre	Post
Quadrilateral	2	10	0	0	2	10
Parallelogram	13	12	17	12	30	24
Rectangle	8	6	7	7	15	13
Square	12	14	9	4	21	18
Rhombus	5	7	1	6	6	13
Trapezoid	5	7	0	0	5	7
Kite	0	1	1	5	1	6

Table 4.10 shows that the word *parallelogram* (n=54) was the most frequently use word during the interviews and it was mentioned in all three tasks. The word *square* (n=39) was the second most frequently used, and the word *rectangle* (n=28) was third. In contrast, the word *kite* (n=6) is the least mentioned during the interviews, used only at the post-interview, in Task 1 (n=1) and Task Two (n=5). There was an increase in use of the word *quadrilateral* at the post-interview, and it was used

mostly in Task One (n=10), and a total of twelve times in the entire interview. Also, there was an increase in use of the word *rhombus* at the post-interview (n=13). The frequencies of using the terms of quadrilaterals provided some information about Molly's familiarity of quadrilaterals, they do not provide details about how and in what way those words were used. Therefore, the analyses of *word use* focus on participants' word meaning in the use of *quadrilateral, trapezoid, kite, parallelogram, rectangle, square,* and *rhombus* during the interviews. In the pre-interview, Molly's use of the words *parallelogram, rectangle, square,* and *rhombus* are visual descriptions of what she observed, as shown in the following dialogue.

> Researcher: What is a parallelogram?
> Molly: A parallelogram is when two sides of each side... all four are
> parallel to the opposite one...
> Researcher: What is a rectangle?
> Molly: A rectangle is the two longer sides... the shorter ones ... they
> are congruent.
> Researcher: What is a square?
> Molly: A square is all four of the sides are completely the same
> Researcher: What is a rhombus?
> Molly: A rhombus... is a square... is just tilted

Molly provided a descriptive narrative about *rectangles* based on their physical appearance, "a rectangles is the two longer sides [and two] shorter ones...they are congruent." Molly made connections between *squares* and *rhombi* according to their visual appearances, and declared narratives, "a rhombus is a square" because "they both have four equal sides," and "[it] is just a tilted [square]." The research pointed to Task One and asked Molly to identify all the parallelograms from a set of given figures.

> Molly: Ok. [Marking stars on figures that are parallelograms]

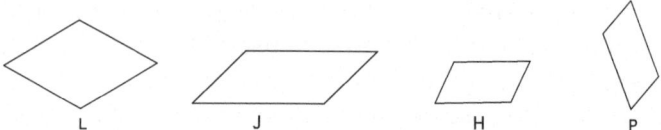

Molly: Now for these ones, these could be actually...be considered parallelograms. Based on the side measures...even though they are rectangles... they could be in the same category.

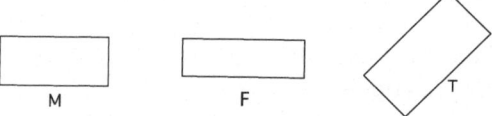

Molly's written narratives about the *rectangles* and *parallelogram* also confirmed what she meant when the terms were used, "rectangle is when 2 [opposite] sides are differing from the other 2 sides, however, opposite sides are equal in length", and "parallelogram is when 2 parallel sides are congruent in length." Molly used the word *parallelogram* as a family name of figures having opposite sides that were parallel, and having two long sides and two short sides. In that case, rectangles "could be considered parallelograms." In her response to Task Two (see Table 4.7) Molly drew two parallelograms: "a stereotypical parallelogram" and a "not typical looking parallelogram." After Molly drew these parallelograms, she was presented a picture of a square and a picture of a rhombus. Molly did not think a square and a rhombus were parallelograms because "to be a parallelogram, you have to have two long sides and two short ones, here all sides are equal and it is square". In the case of a rhombus, Molly responded, "this is similar to the square that you just showed me, ...is a rhombus or just a square". The following diagram illustrates Molly's use of word parallelograms at the pre-interview.

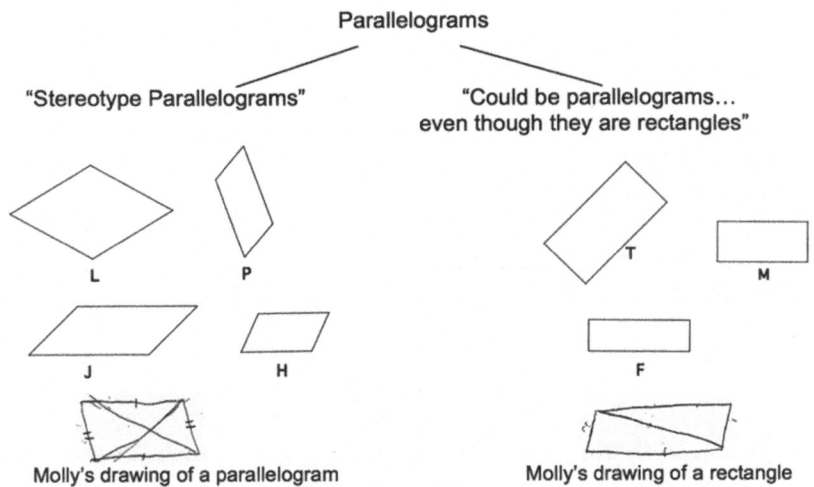

Figure 4.7. Molly's use of the word parallelogram in the pre-interview.

Molly's use of the word *parallelogram* signified a collection of unstructured polygons by their family appearances. It included figures appearing to have opposite sides equal and parallel, and in particular, two opposite sides longer than the other two opposite sides. However, there was no explicit mention of the necessary condition that these figures be 4-sided, nor of any condition on the angles in rectangles.

At the post-interview, Molly grouped the quadrilaterals into: (1) *squares* (U, G, and R), (2) *rectangles* (M, F, and T); (3) *rhombi* (Z); and (4) *parallelograms* (L, J, and H). She explained her grouping as: "Quadrilaterals, you have your square because each form 90-degree [angles] and all side lengths are equal [U]. These are rectangles [F and M] because those two sides are the same. But again they form 90-degree angles. Opposite angles are equal and opposite sides are equal, so these three would be an example of parallelogram [L, J and H]."

Molly did not spell out exact mathematical definitions of squares, rectangles, and parallelograms, but she was able to group them together based on the common features relating to angles and sides that she observed. To investigate further, the researcher asked Molly whether J and Z, and U and M, could be grouped together.

Molly: Mm Hmm [yes].

Researcher: Why is that?

Molly: [...] because they both have opposite sides parallel and opposite angles are equal.

Researcher: Can I group figures U and M together?

Molly: Yeah, you can because U has the same property as M. The only difference is that M does not have all the same sides length, so M would not have all the properties as U, but U has all the properties of M.

Molly recognized that a rhombus is also a parallelogram, and noted that she did not initially group a rhombus and a parallelogram together. She also agreed that a square and a rectangle could be grouped together, because a square shares a property with a rectangle. In both cases, when prompted, Molly agreed that some quadrilaterals could be grouped together, but she did not on her own assign a common name (e.g., "rectangle") to the group. When Molly was asked to identify *all* the parallelograms among the quadrilaterals, she replied, "L, J and H will be *just* parallelograms, but all of these figures [pointing to squares, rectangles and rhombi] could be parallelograms, because they all fit to the greater property of "opposite angles and sides to be equal." It is not clear why figure P (another parallelogram) was missed from her grouping.

When requested, Molly provided her of definitions as shown in Table 4.11 for square, rectangle, parallelogram and rhombus.

Table 4.11. Molly's Definitions of Parallelograms in the Post-Interview

Name	Definitions	Sample Polygons
Square	A Square is when all the angles form right angles and they are all the same they are all 90 degrees and each side length also has to be the same [pointing at U]	U
Rectangle	A Rectangle, each angle is 90 degrees but these sides are the same and parallel, and this one is the same and parallel, but not all 4 of them are the same, necessarily [pointing at M]	M
Parallelogram	A parallelogram is when opposite sides are equal and opposite angles are both equal [pointing at J]	J
Rhombus	As rhombus alone, it has all sides the same, but does not form 90-degree angle [pointing at Z]	Z

Molly's use of the word parallelogram signified a family of quadrilaterals that shares this common description: "they all fit to the greater property of opposite angles and sides to be equal." The following diagram is an illustration of Molly's use of the *parallelogram* at the post interview.

In these conversations, more dimensions were added to Molly's use of the word *parallelogram*. At the pre-interview, the word *parallelogram* only signified polygons that fit into the physical appearances of parallelograms and rectangles whereas at the post-interview, the word *parallelogram* signified a family of polygons that share common descriptive narratives.

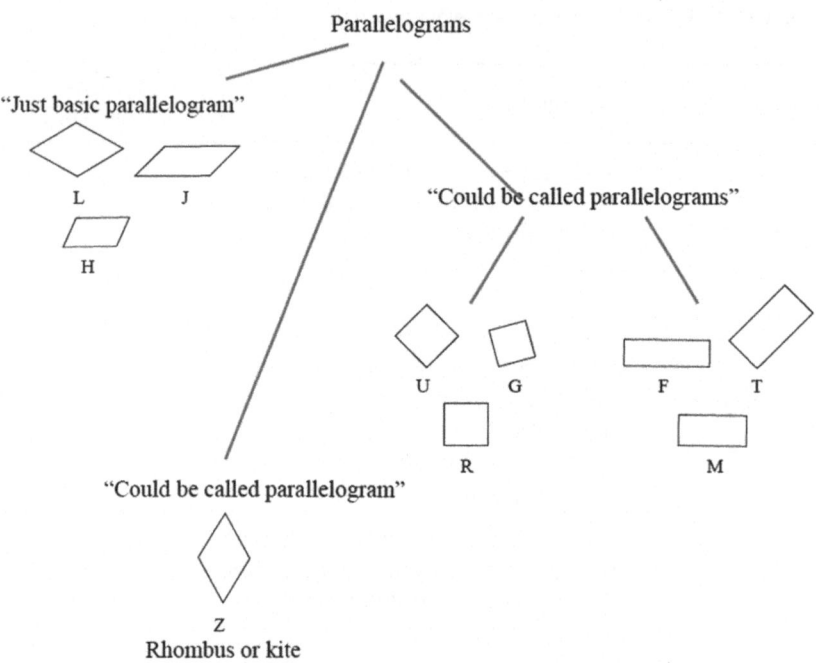

Figure 4.8. Molly's use of the word parallelogram in the post-interview.

As shown in Figure 4.8, the word *parallelogram* signified to Molly a common family name for all figures that "have opposite sides parallel and opposite angles equal." This diagram illustrates how parallelograms were inter-connected. For example, Molly identified that "as a rhombus alone" [it] does not form a 90-degree angle, and "sides are all the same." A rhombus was different from a square with regard to the angles: "all the angles form right angles…and each side length also has to be the same." However Molly did not mention how rectangles were different from parallelograms.

4.2.2   Case Two: Changes in Ivy's Geometric Discourse

Ivy was a second-year college student and took her last geometry class six years prior to the study. The van Hiele geometry Tests showed that she was at Level 0 at the pre-test, and moved to Level 3 at the post-test. Ivy's changes in geometric discourse from the pre-interview to the post-interview are summarized as follows:

- Ivy's routines of sorting changed from grouping by the names of polygons, according to their family appearances and visual properties with no order, to classifying polygons according to their common descriptive narratives, and structuring quadrilaterals with a hierarchy of classification.
- Ivy's routines of substantiation changed from visual recognition and recalling, to using endorsed narratives such as definitions and properties of parallelograms and from comparing parts of parallelograms visually to applying various methods (e.g., Pythagorean theorem, congruence criterion, algebraic derivations) to verify claims at meta-level.
- Ivy's use of the names of parallelograms changed from visual recognition of their family appearances to using these names as collections of quadrilaterals sharing common descriptive narratives in a hierarchy of classification.

Changes in Ivy's Routines
During the pre-interview, Ivy first sorted polygons by their names based on family appearances, finding *triangles* (S, X, W, K and V), *squares* (U, G and R), *rectangles* (M, F and T), *parallelograms* (L, Z, P, J and H), and *trapezoids* (N and Q). She grouped a hexagon (V) with the triangles because she dissected two triangles in it.

Ivy: I made a group of triangles, which had three sides...

Ivy: I included V as two triangles, because if you draw a line here, it would make two.

(an illustration of what Ivy meant)

Ivy grouped a right trapezoid (N) and a quadrilateral (Q) together as a *trapezoid* group because she was not sure about what to do with these two polygons.

Ivy: I did one that just had Q and N, which I didn't know what to classify. I guess they would be trapezoids.

When asked for regrouping, Ivy combined rectangles and squares together in one group because "every square is a rectangle." She then split the triangle group into right triangles and isosceles triangles, but did not know what to do with two other triangles.

Ivy: That's a right triangle [K], and that's an isosceles triangle [W].

Ivy: I don't know what name is for [X], and this one [S], none of the sides on that looked even to me, so I don't think it looked like any of the others

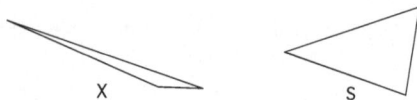

Ivy grouped two triangles according to the visual properties of angles (i.e., right triangle) and sides (i.e, isosceles triangle), and left two other

triangles (X and S) with no groups. Ivy explained her subgrouping of the parallelograms (L, Z, P, J and H).

> Ivy: L and Z look more like square, I don't know if there are two different types of parallelograms.

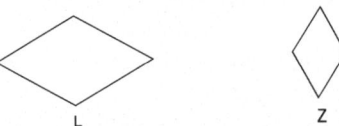

> Ivy: Where P, J, and H look more of a rectangle, like the opposite side of all of these look the same. So, I guess you could put it into two groups that way.

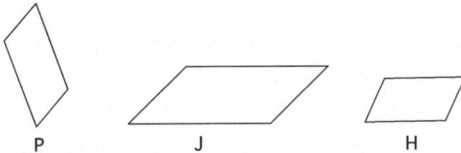

Ivy split the parallelograms into two groups: *squares* (L and Z), and *rectangles* (P, J and H), by their visual appearance. Ivy's definitions of *parallelogram, rectangle, rhombus* and *square* are summarized in Table 4.12.

Table 4.12. Ivy's Definitions of Parallelograms in the Pre-Interview

Name	Definitions	Sample Polygons
Square	Object with four right angles, all sides are equal length and opposite sides are parallel.	
Rectangle	Object with four right angles, opposite sides are parallel and equal in length.	

Parallelogram	Objects with four sides. Opposite sides are the same length and parallel. Opposite angles also equal. I guess it's kind of like a slanted rectangle.	
Rhombus	Object with four angles of the same measure, opposite sides are parallel. It is like a slanted square and a parallelogram is a slanted rectangle.	

Ivy described a square as having four sides of the same length, whereas a rhombus was a slanted square. However, she did not mention right angles, a defining condition of rectangles and squares among the parallelograms. The researcher asked to find out whether Ivy considered squares as parallelograms.

Researcher: Can I group J and U together?

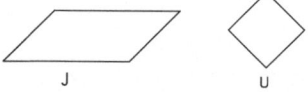

Ivy: You could if you talked about angles, I guess. You'd say this angle and this angle are equal [opposite angles of J], and this angle and this angle are equal, where the opposite angles are equal here [U]. So, I guess you could group it in that way.

Ivy agreed a parallelogram (J) could group with a square (U) because opposite angles were equal in both polygons. Ivy did not identify a square as a parallelogram because "a square has all the same length, and a parallelogram has different sides." In these actions, Ivy's identifying routines focused on visual property of their angles. When the interviewer asked Ivy if a parallelogram (J) and a rectangle (M) could group together, her response was different.

> Ivy: Yeah. You also have opposite sides are parallel [pointing at the opposite sides of J]...and this side and this side are both shorter than the other ones [pointing at the opposite sides of M], they [J and M] have quite a bit in common.

Ivy acted more positive towards the grouping of a parallelogram (J) and a rectangle (M), and thought, "they have quite a bit in common." Ivy explained that a rectangle was a parallelogram "because it has two opposite sides equal, a rectangle and a parallelogram go together." In these actions, Ivy identified polygons by visual property of their sides. Ivy did not use any measurement tool to check the parts of polygons in the task. Her routines of regrouping polygons included direct recognition and identification of polygons by visual properties of angles and sides. Ivy's routines of sorting polygons at the pre-interview are summarized in Figure 4.9.

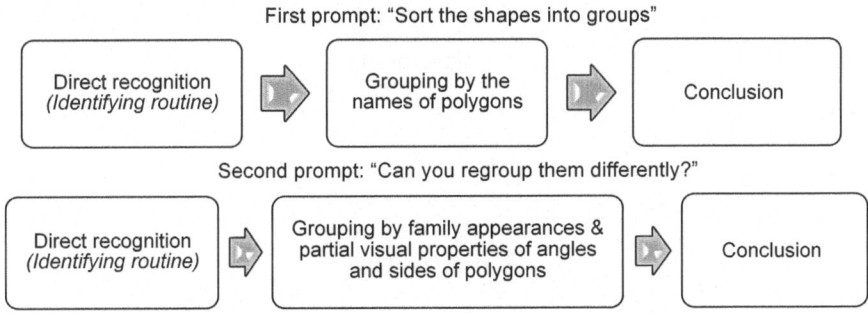

Figure 4.9. Ivy's routines of sorting polygons in the pre-interview.

Ten weeks later, Ivy's routines of sorting polygons changed from grouping polygons by their visual appearance, to classifying them by common descriptive narratives with a hierarchy of classifications.

At the post-interview, Ivy first grouped polygons by their names, finding *quadrilaterals* (U, M, F, G, P, N, L, J, Z, Q, T, H, and R), *trapezoids* (N), *parallelograms* (P, H, U, G, T, R, M, F and J), *rectangles* (M, F, T, U, R and G), *rhombi* (U, G, R, Z and L) and *triangles* (K, W, X and S). Ivy grouped all 4-sided figures into the *quadrilaterals* group, and all 3-sided figures into the *triangles* group. She included parallelograms, squares, and rectangles as *parallelograms*, but not the rhombi. The *rectangles* group consisted of rectangles and squares, and the rhombi group included squares and rhombi. Note that a hexagon (V) was not included in any of these groups. As an example, Figure 4.10 presents Ivy's groups of *parallelograms*, *rectangles* and *rhombi*.

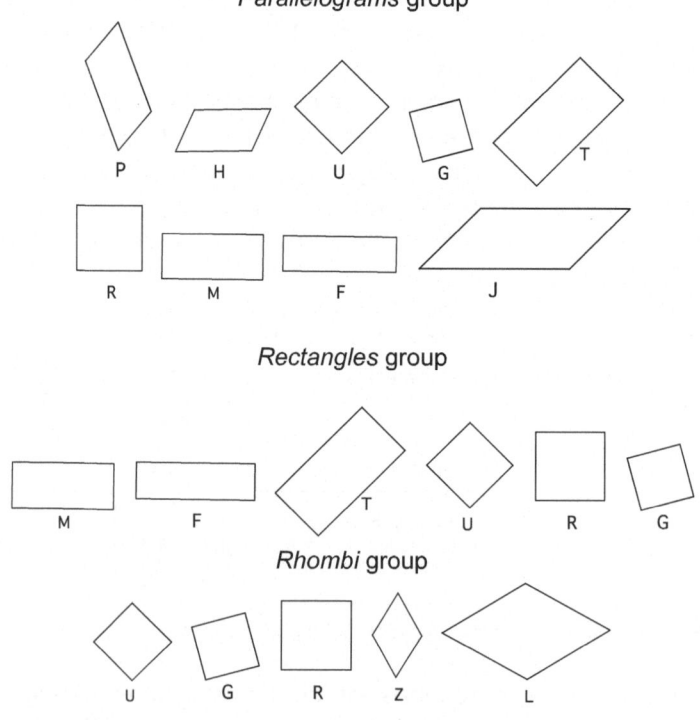

Figure 4.10. Ivy's grouping of parallelograms in the post-interview.

Figure 4.10 illustrates Ivy's routines of grouping polygons focused on identifying their common descriptive narratives (i.e., definitions). For example, Ivy explained that she grouped squares and rectangles together as a *rectangles* group because "all squares are rectangles, and rectangles have all 90-degree angles." She verified that both squares and rhombi were rhombi because "that is what a rhombus is, four sides of equal length." More importantly, Ivy demonstrated the ability of classifying the quadrilaterals within a hierarchy. Such classification includes *trapezoid*, a quadrilateral with one pair of parallel sides; *parallelograms*, quadrilaterals with two sets of parallel sides; rhombus, quadrilaterals with all sides equal; and a quadrilateral with no two sides equal. Ivy did not use rulers or protractors to check measurements of angles and sides at the post-interview, but she did mention that her conclusions were made under the assumptions of the sides looked like they were parallel, or angles looked like they were right angles, etc. Figure. 4.11 Summarizes Ivy's routines of sorting polygons in the post-interview.

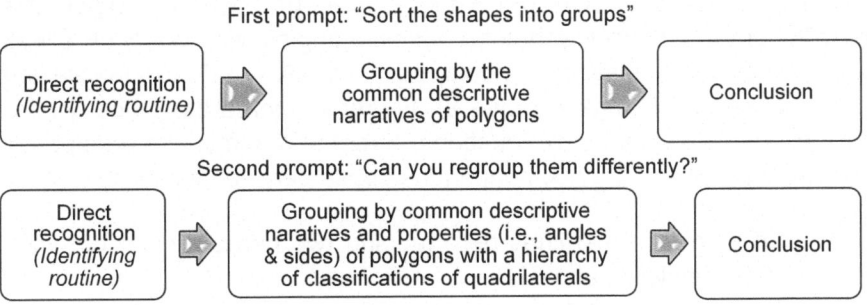

Figure 4.11. Ivy's routines of sorting polygons in the post-interview.

Ivy also demonstrated her ability to conduct geometric proofs during the post-interview using substantiation routines, which was absent in the pre-interview. In particular, she used triangle congruence criterion to substantiate her claims about the properties of parallelogram. She also used the Pythagorean Theorem to derive some statements algebraically at the post-interview. For example, in the pre-interview, she drew a parallelogram and labeled the vertices as A, B, C and D in clockwise order. She then wrote a statement regarding the angles of the parallelogram, "∠A = ∠C, ∠B = ∠D". The following conversation took place when Ivy was asked for verification.

Researcher: What can you say about the angles of this parallelogram?

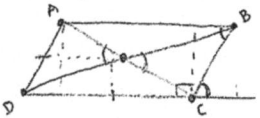

Ivy's drawing of a parallelogram

Ivy: Angles A and C are equal, and then B and D are equal [Pointing to angles A and C, she wrote: ∠A = ∠C, ∠B = ∠D].
Researcher: How do you know?
Ivy: I just remember being taught that, I don't actually know...If you measured them, they would be equal.

In her response, Ivy referred the prior knowledge about the angles of a parallelogram to conclude that the opposite angles were equal. She was able to use mathematical symbols, such as ∠A = ∠C, to indicate the equivalence of the angles. However, Ivy remembered the property as a fact without knowing the explanations. After the researcher's prompt, Ivy verified that ∠A = ∠C by comparing the space between the angles.

Ivy: If you drew two lines here [adding two perpendicular dot lines from angles A and C]

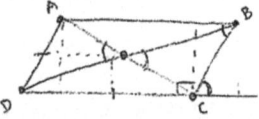

Ivy: From this line to this line, if you know that's a 90-degree angle...[pointing at the indicated space between AB and perpendicular lines are formed]

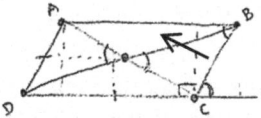

Ivy: You know what a 90-degree angle looks like, so it's easier to go off of that and then you look at the space here and the space here [pointing at the marked angles], and see if those are the same.

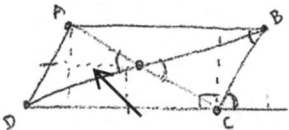

Ivy assumed ∠A = ∠C because the two angles were the sum of a right angle and a smaller angle. She started with comparing the right angles because they were easy to distinguish by their visual appearance and then she compared *space* between the two smaller angles to check whether they were the same. These actions show that Ivy's routine procedure relied on the visual appearance of the angles to verify her claims.

Ivy made two other statements about the angles of the parallelogram, "∠A + ∠D = 180°, ∠C + ∠B = 180°". When asked for substantiation, she responded as follows.

Ivy: If you extend this line out, and draw another straight line there...

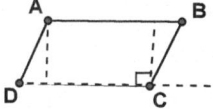 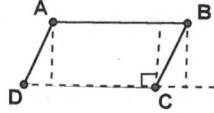

An illustration of what Ivy meant

Ivy: You can see that this angle equals that angle.

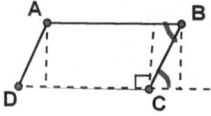

Ivy pointing at the two marked angles

Researcher: How do you know they are equal?

Ivy: There is a term for it. It's like a rule that I remember learning. Maybe parallel angle rule? Adjacent angle rule? Or something.

Ivy: And this is angle C [pointing at the ∠BCD]. So if you were to combine them, you would have a straight line, and that would make a 180-degee angle.

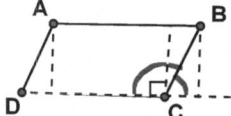

Ivy pointing at the two marked angles

Ivy recognized the structure of alternating interior angles formed by parallel lines and their transversals. When she explained, "you can *see* that this angle equals that angle" she again relied on the visual appearance of the angles. She did not know names of the angles, nor the related propositions to support her claim, but referred to a rule that she had learned. Assuming that the two marked angles were equal, Ivy verified that ∠C + ∠B = 180° because they made a 180-degree angle. Using the same reasoning, Ivy also stated that "∠D + ∠A = 180°". Ivy used visual recognition to compare angles, and applied prior knowledge as a fact to verify the statements about the angles of a parallelogram.

In the post-interview, Ivy drew a parallelogram and labeled the vertices as A, B, C, and D in a clockwise rotation, and gave the same statements about the angles of the parallelogram as in the pre-interview. Figure 4.12 shows Ivy responses for "What can you say about the angles of this parallelogram?" in the post-interview.

A. Draw a <u>parallelogram</u> in the space below.

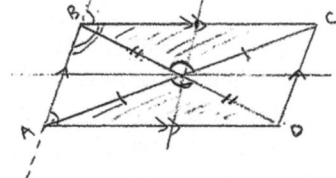

1. What can you say about the angles of this parallelogram?

$\angle A = \angle C$ $\angle A + \angle B = 180°$ $\angle C + \angle B = 180°$

$\angle B = \angle D$ $\angle D + \angle C = 180°$ $\angle D + \angle A = 180°$

Figure 4.12. Ivy's written responses for Task 2 in the post-interview.

In contrast to her responses in the pre-interview, Ivy's verification of her claims was different. When asked for substantiation of the statements ∠A = ∠C and ∠B = ∠D, she explained that "this is a parallelogram because I drew it, so ∠A is equal to ∠C and ∠ B is equal to ∠D". Ivy's verification of ∠A + ∠B = 180° is shown next.

Ivy: If you were to extend this line (AB)…

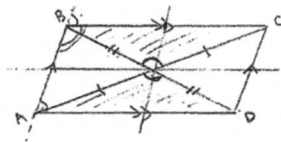

Ivy: You could look either way, like this angle is equal to this angle. BC and AD are parallel. They [∠A and it corresponding angle] are corresponding angles because they are on the parallel lines.

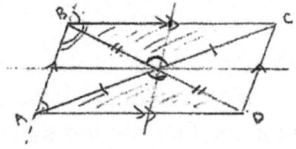

Ivy: Then you could tell that if you add these two angles [∠ABC and it's adjacent angle], it's angles on a line. So it's 180 degrees.

Ivy: So angle A and angle B add up to 180 degrees.

To verify ∠A + ∠B = 180°, Ivy extended side AB so that the structure of the corresponding angles formed by parallel lines and their transversals was visible. She mentioned the corresponding angles were congruent *because* they were on the parallel lines, and then concluded that ∠ A and ∠ B add up to 180 degrees. Although Ivy verified her claim informally, it is important to see the change in her deductive reasoning: Ivy justified her claim that corresponding angles were equal using an endorsed narrative that BC is parallel to AD at the post-interview, whereas her reasoning relied on visual appearances of lines and the angles at the pre-interview.

Ivy provided different narratives about the diagonals of a parallelogram at the two interviews. At the pre-interview, she asserted that the diagonals of a rhombus should be the same and the diagonals of a rectangle should intersect at the middle of the rectangle, whereas at the post-interview she stated that the diagonals of a rectangle should be equal in length, and the diagonals of a parallelogram bisect each other at the post-interview. In the following, the results of Ivy's routine procedures of substantiating these narratives at both interviews will be described and compared.

In the pre-interview, Ivy drew a rhombus, and stated that it was a parallelogram and it was different because "all the sides were the same length." When asked for the angles and diagonals of the parallelogram (i.e., rhombus), her responses are as follows.

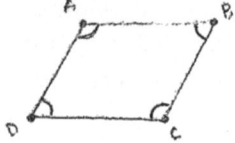

Ivy's drawing of a rhombus

Ivy: I think they [the angles] should all the same. I guess if I had drawn it better, all the angles should be the same.
Researcher: Why do you think they are the same?

Ivy: because the lengths [the sides of the rhombus] are the same. [Wrote: ∠A = ∠B = ∠C = ∠D because lengths of the sides are the same]

Researcher: What can you say about the diagonals of the parallelogram?

Ivy: I think the diagonal should be equal in length.

Researcher: How do you know they should be equal in length?

Ivy: Because the sides are all the same length and the angles are all the same. So, I think the diagonals should be the same.

In the preceding conversation, Ivy made incorrect claims about the angles and diagonals in this rhombus (a rhombus without right angles), saying they, "should *all* be the same." When asked for verification, she explained that all the angles should be the same *because all the sides of a rhombus were the same* and diagonals should be equal in length *because all the sides and angles of a rhombus were all the same*. For Ivy's conclusions about the angles and diagonals of this rhombus to be correct, the rhombus had to be a square (a rhombus with right angles). Ivy made incorrect implications from the equivalence of the sides to the equivalence of the angles, and then suggested that the diagonals must be equal because of equal sides and angles. There is no routine involved in this verification, other than making statements based on the fact that all the sides are equal in a rhombus. Ivy demonstrated an incorrect understanding of a rhombus at the pre-interview.

In the case of a rectangle, Ivy stated that the diagonals of a rectangle were longer than its sides, and the intersection of the diagonals was at the middle of the rectangle.

Ivy: The length of this [pointing at the diagonal] is longer than the length of the longest side [pointing at the longer side of the rectangle].

 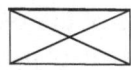

Ivy: They [the diagonals] should intersect in the middle

Ivy declared two narratives about the diagonals of this rectangle based on visual recognition of what the diagonals appeared to be. She recognized the diagonal as the hypotenuse of a right triangle, and mentioned "the Pythagorean Theorem, which is how I know it" to verify the diagonals were longer than the longest side of rectangle. Her argument that the diagonals should be at the middle of the rectangle went as follows:

Ivy: So, if you were to find the midpoint of this length… [Drew one line passing through the midpoint of the sides]

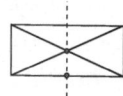

Ivy: …and then if you were to find the midpoint of this length, and draw a line [Drew another line passing through the midpoint of the other sides]

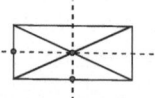

Ivy: … that the intersection of those two lines should be the intersection of the diagonals as well.

Ivy verified that the diagonals intersect at the middle of the rectangle by locating the midpoints of the sides of the rectangle, and concluded that the intersection of the two medians was the same point as the intersection point of the diagonals.

In the preceding examples, Ivy used the Pythagorean Theorem to conclude that the diagonals of the rectangle are longer at its sides. Her understanding of the properties of parallelograms (i.e. rectangle) was not clearly demonstrated. To verify her claims, she focused visual recognition intuitively at the pre-interview.

At the post-interview, Ivy drew a parallelogram, and stated that the diagonals of the parallelogram "are not equal in length" and "the diagonals bisect each other creating corresponding triangles."

Ivy: They [the diagonals] cross at one point.

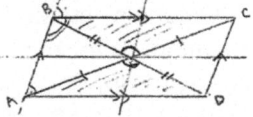

Ivy's drawing of a parallelogram

Ivy: They create corresponding triangles. Well, like this triangle corresponds with this triangle [Shaded the two corresponding triangles]

Researcher: What do you mean by "corresponding triangles?"
Ivy: This angle and this angle are equal, cause they're vertical angles [marked angles at the intersection of the diagonals]

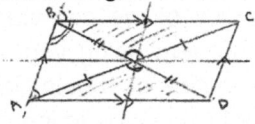

Ivy: And then, this side should equal this side [marked corresponding congruent sides cut by the diagonals]. And this side should equal this side. And I know they're corresponding [triangles] through Side-Angle-Side.

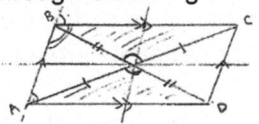

Researcher: How do you know these corresponding sides are equal?
Ivy: because the diagonals bisect each other.
Researcher: How do you know they [diagonals] bisect each other?

> Ivy: I don't really know. I guess it's because the sides [of the parallelogram] are equal and they're parallel, so…

In the preceding conversation, Ivy started with a descriptive narrative about the diagonals of a parallelogram, saying "they cross at one point," and then she asserted that the diagonals created corresponding triangles. When asked for substantiation, she responded that the corresponding triangles [shaded in Ivy's drawing] were a pair of congruent triangles using the Side-Angle-Side (SAS) criterion. Ivy used an endorsed narrative "diagonals bisect each other" to imply the corresponding sides of the triangles were congruent. However, when asked how she knew the diagonals of this parallelogram bisect each other (a property of a parallelogram), she responded, "I don't really know…I guess, it's because the sides are equal length and they're parallel."

Ivy used her substantiation routines to conduct an informal proof using SAS congruent criterion. That is, she identified corresponding triangles, and three elements needed for verification of congruent triangles: two sides and an included angle. She used "diagonals bisect each other" to justify the congruency of the corresponding sides, and showed included angles were congruent using vertical angle theorem. Thus, there was a change in Ivy's routine of substantiating, from intuitive visual recognition to using endorsed narratives to identify three elements for verifying congruent triangles at the post-interview.

When discussing the diagonals of a rectangle, Ivy provided a narrative "diagonals bisect each other and they are equal in length," and tried to verify the claim using the Pythagorean Theorem. She first labeled diagonals with four segments a, b, c and d, and wrote "a + b = c + d" to indicate the diagonals of the rectangle are equal.

> Ivy: You could find that a is equal to d, they should be equal.

=>

As an illustration

Ivy: I could use the Pythagorean Theorem again, but with this side and this side. And then I'd find *d*.

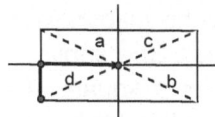

Pointing at the two legs of the right triangle

Ivy: *a* equal to *d* because they share this side here [pointing at the longer leg], and this point is a middle point here, so these two sides are equal [two shorter legs]. They [*a* and *d*] are equal.

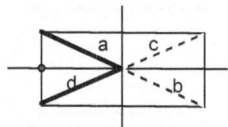

Ivy provided correct conclusions about the diagonals of the rectangle "diagonals bisect each other and they are equal in length." When asked for substantiation, Ivy described informally her verification of the claim "*a* = *d*" by identifying the two triangles sharing a longer side and having equal shorter sides and by applying the Pythagorean theorem to conclude "*a* = *d*". Ivy did not show the two triangles were right triangles, an important condition of using the Pythagorean Theorem, nor did she give details of the algebraic derivation of "*a* = *d*". However, there was a change in Ivy's routine of substantiation provided she was able to derive partial informal proof of "*a* = *d*" under the assumption that the two triangles are right triangles using Pythagorean Theorem in the post-interview.

In the case of square (a rhombus with right angles), she produced two narratives about the diagonals of the square in the post interview.

Researcher: What can you say about the diagonals of the square?

Ivy: They're equal… They bisect the angles, split the angles into two-45 degree angles.

Researcher: How do you know they are equal?

Ivy: The same way I knew with the rectangles.

Researcher: How do you know they [diagonals] bisect the angles?

Ivy: It [diagonal] divides the angle into two equal angles.

Ivy provided two narratives about the diagonals of the square, "they're equal," and "they bisect the angles." She applied her knowledge of the diagonals in a rectangle to case of a square. To verify the diagonals bisect each other, Ivy explained that they divide the angle into two equal angles. Table 4.13 illustrates Ivy's verification routines with corresponding transcripts.

Table 4.13. Ivy's Verification Routines of "Diagonals Bisect the Angles" in the Post-Interview

Routine Procedures	Transcripts
1. Declare narratives	
1.1 Draw a diagonal	I guess I'd draw a diagonal
1.2 Identify two right triangles	It splits the square into two right triangles, because all of these angles are 90-degrees. *adding right angle sign on each angle of the square*

1.3 Identify the relation between the angles and sides of the right triangle.	174c. By the angle sum rule, all angles add up to 180 degrees. You already have 90 here. So, X plus Y has to equal 90. It's also an isosceles triangle. *assigning X and Y to the two angles*
	Q: How do you know it's an isosceles triangle?
2. Verification of isosceles triangle 2.1 Identify congruent sides of the triangle	180. These two sides equal. *Adding two marks on the sides of the triangle*
2.2 Verification of congruent angles	182a. It's an isosceles triangle. So X is equal Y.
2.3 Finding the angle measures of X and Y	182b. I know that X and Y has to equal 90 degrees. So, I know that X is 45 degrees and Y is 45 degrees.
2.4 Finding other angles measures	190. So, if you know it's 90, and Y is equal 45 degrees, and this angle is also 45 degrees. Same for X here. So diagonals splitting into two equal angles and they are 45 degrees each. *"this angle is also 45 degrees"*
3. Conclusion	194. Yeah, diagonals bisect each other

It became clear that Ivy favored algebraic reasoning in her substantiation routines. She labeled the angles X and Y and used an endorsed narrative, "all angles are 90 degrees" to justify that X and Y were the angles of a right triangle. She used another endorsed narrative, "these two sides equal" to verify that the triangle is isosceles. Finally, Ivy solved X and Y algebraically, to find that they were 45 degrees each. Using this newly endorsed narrative, Ivy concluded that the diagonals bisect the angles [of the square]. These examples demonstrate the changes in Ivy's routines at the post-interviews. In particular, Ivy used her knowledge in algebra to help solve problems in geometry.

Changes in Ivy's Word Use
Ivy's use of the word *parallelogram* changed from describing the visual appearances of the quadrilaterals at the pre-interview, to using the word as a common descriptive narrative with a hierarchy of classifications in the post-interview. The frequencies of Ivy's use of the names of quadrilaterals in the two interviews are presented in Table 4.14.

Table 4.14. Frequencies of Ivy's Use of the Names of Quadrilaterals in the Two Interviews

Name	Frequency				Total Frequency	
	Pre-T1	Post-T1	Pre-T2	Post-T2	Pre	Post
Quadrilateral	0	3	0	0	0	3
Parallelogram	4	3	4	0	8	3
Rectangle	12	3	2	2	14	5
Square	10	5	1	5	11	10
Rhombus	2	2	1	3	3	5
Trapezoid	2	2	0	0	2	2
Kite	0	1	0	0	0	1

Table 4.14 shows the word *square* (n=21) was the most frequently used during the interviews. The word *rectangle* (n=19) was the second most frequently used, and *parallelogram* (n=11) was third. The names of the parallelograms were mostly mentioned in Task One, and among all the names, *rectangle* and *square* were most frequently mentioned at the pre-interview. There was an increase in use of the word *square* and *rhombus* in the post-interview. However, there was a reduction in use of the word *parallelogram* and *rectangle* in the post-interview. The words

kite (n=1) and *quadrilateral* (n=3) were mentioned only at the post-interview, whereas the word *trapezoid* (n=4) was only mentioned in Task One. Ivy's use of the names of quadrilaterals was much lower than other interviewees' use of those names.

Recall that Ivy grouped the parallelograms as parallelograms (P, L, Z, H, and J), and rectangles (M, F and T) at the pre-interview. In particular, the parallelograms also included rhombi (L and Z). While grouping polygons, Ivy referred to a rhombus as a slanted square, and a parallelogram as a slanted rectangle. Later she drew a rhombus as a parallelogram, but disagreed that a square was a parallelogram when she was presented a picture of a square.

Researcher: Why is this a parallelogram?

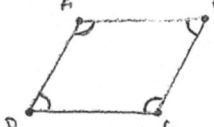
Ivy's drawing of a rhombus

Ivy: Because AB is parallel to DC, and AD is parallel to BC.
Researcher: Why do you think it's a different parallelogram?
Ivy: Because all the sides are the same length.
Researcher: What do you call this shape?
Ivy: A rhombus.

Ivy identified a rhombus as a parallelogram because it had two pairs of parallel sides and recognized that it was a different parallelogram because it had all sides of the same length. However, in the next conversation, she disqualified a square as a parallelogram.

Researcher: How about this one? Is this a parallelogram?

Ivy: No.
Researcher: Why do you think it's not a parallelogram?

Ivy: Because all the lengths look like they are the same sides, and I
    think that a parallelogram has different sides.
Researcher: What do you call this shape?
Ivy: A Square.

This inconsistency showed that Ivy's use of the word *parallelogram*
referred to visual family appearances. Ivy's grouping of parallelograms
and her ways of identifying and defining parallelograms showed that she
used the word *parallelogram* to represent *rectangles, parallelograms* and
rhombi. Figure 4.13 illustrates Ivy's use of the word parallelogram.

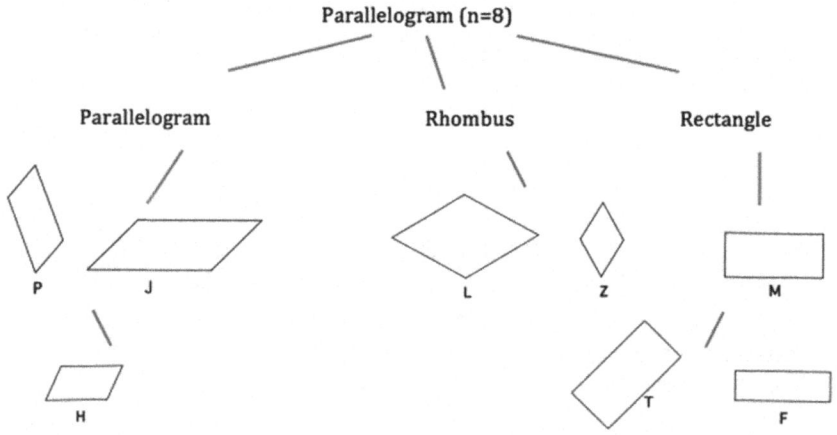

Figure 4.13. Ivy's use of the word parallelogram in the pre-interview.

Ivy's use of the word parallelogram changed in the post-interview.
During the interview, She grouped all 4-sided figures (n=13) into the
quadrilaterals group, which includes rectangles, parallelograms and
squares as *parallelograms*, but not all the rhombi. The *rectangles* group
consisted of rectangles and squares, and the rhombi group included
squares and rhombi. Figure 4.14 presents Ivy's groupings of
parallelograms, rhombi and rectangles.

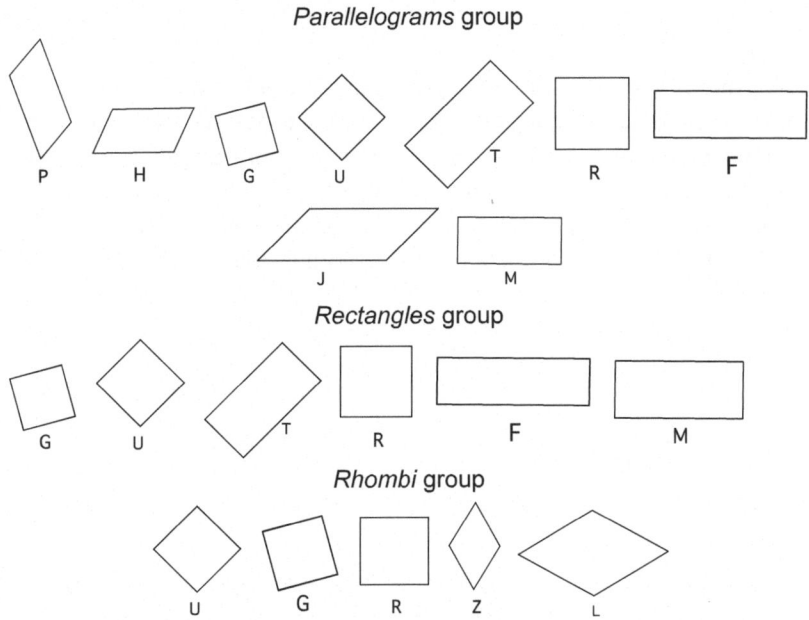

Figure 4.14. Ivy's grouping of parallelograms in the post-interview.

Ivy explained, "parallelogram [J] is two sets of opposite parallel sides, and I did the ones that are kind obvious, I also did the ones that you'd think of as a rectangle [F] and a square [G] and rectangle are the ones that look like they had 90-degree angles, and I included squares cause squares are always rectangles," and then she continued, "rhombi have four equal sides and I have Z, R, G, U, and L", and verified that both squares and rhombi were rhombi because "that is what a rhombus is, four sides of equal length." To justify her responses in grouping rectangles and parallelograms together, Ivy also argued, "they're parallelograms because they have two sets of opposite sides parallel."

Ivy used a definition of parallelogram to group all qualified quadrilaterals, and classified the quadrilaterals beginning with the attributes of their sides. When asked to create a different grouping, Ivy split her parallelograms group into parallelograms and rectangles, and split the rhombus group into rhombi and squares by the characteristics of right angles. Figure 4.15 is generated to illustrate Ivy's use of the *parallelogram*.

Ivy's use of the word *parallelogram* signified a hierarchical (and nested) classification of quadrilaterals. To her, these parallelograms may have different visual appearances—some have right angles and some do not—but they all share a common descriptive narrative: "opposite sides parallel and equal." That is, Ivy focused on using a definition of parallelogram to identify a collection of quadrilaterals.

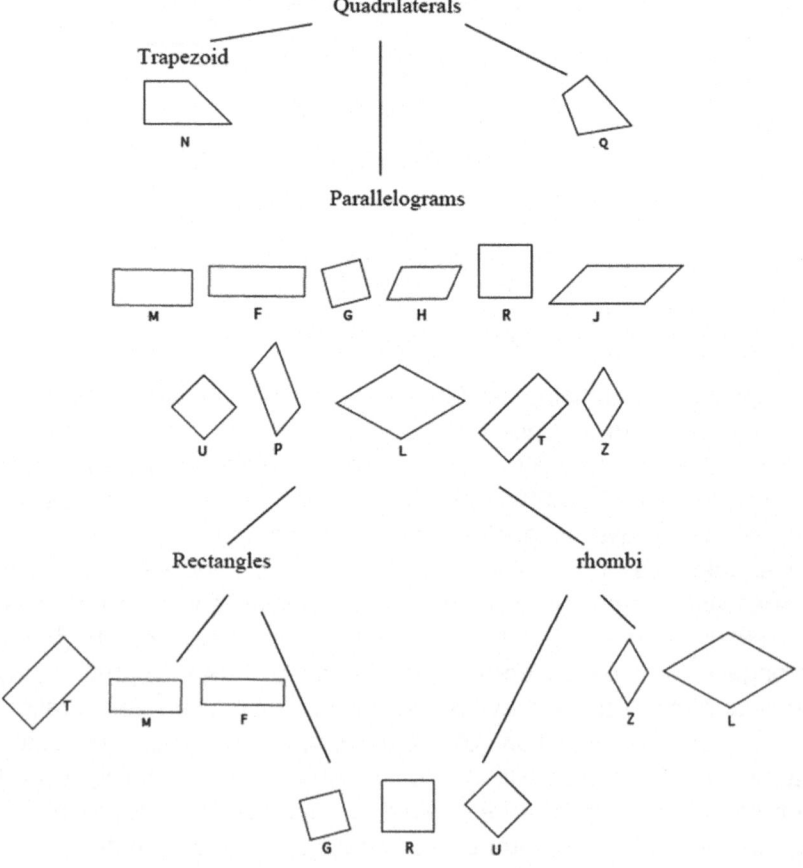

Figure 4.15. An illustration of Ivy's use of the word *parallelogram*.

4.2.3 Case Three: Changes in Kevin's Geometric Discourse

Kevin was a first-year college student at the time of the interviews. He took his last geometry class three years prior to the study. The van Hiele geometry test showed that he was at Level 3 at the pre-test, and moved to Level 4 on the post-test. Kevin was one of the two students who reached Level 4 in the study. Kevin was interviewed after both tests and a summary of the results on changes in Kevin's geometric discourse are as follows from the pre-test to the post-test:

- Kevin's routines of sorting changed from grouping polygons by the number of sides and by their names and angles, to classifying polygons by their common descriptive narratives and arranging quadrilaterals with a hierarchy of classifications.
- Kevin's routines of substantiation changed from verifying the congruent parts of parallelograms using recalling, measuring and constructing routines at the pre-interview, to formulating mathematical proofs using mathematical axioms and propositions.
- Kevin's use of the word parallelograms also changed from using the words as a collection of quadrilaterals sharing common descriptive narratives, to using them with a hierarchy of classifications of parallelograms.

Changes in Kevin's Routines
Kevin classified eighteen polygons into *triangles* (K, W, S and X), *quadrilaterals* (U, M, F, G, P, N, T, L, J, Z, Q, H and R) and *a hexagon* (V). When asked for regrouping, his first approach was to regroup quadrilaterals into *rectangles* (U, M, F, G, T and R) and *non-rectangles* (P, N, L, J, Z, Q and H). He responded, "I selected shapes with right angles… from looking at it. I didn't measure any of them, but I am assuming that they are right angles." Kevin's grouping of rectangles and non-rectangles is shown in Figure 4.16.

Rectangle Group (n=6)

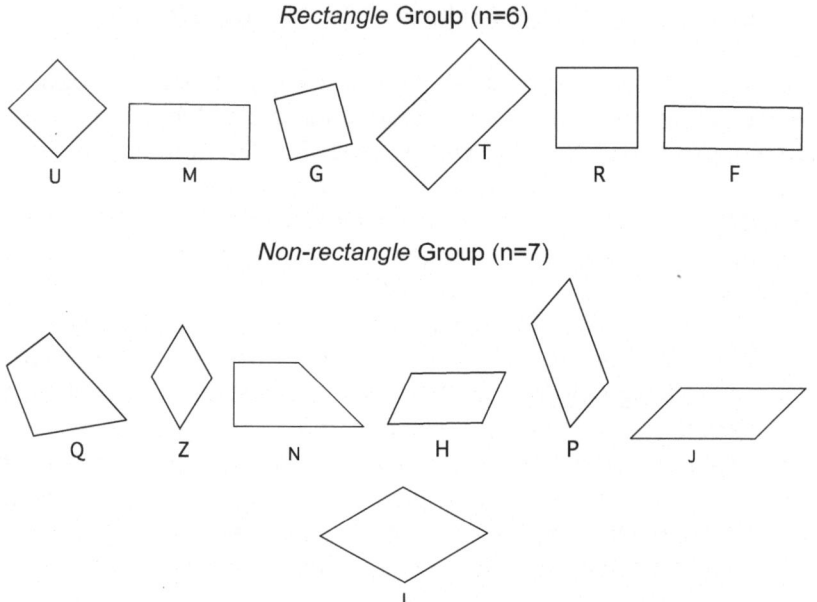

Non-rectangle Group (n=7)

Figure 4.16. Kevin's subgrouping of the quadrilaterals in the pre-interview.

Then, Kevin regrouped eighteen polygons into two groups according to their angles: *right angles* and *non-right angles* (Figure 4.17). He split the *non-right angle* group into two subgroups: one contains obtuse angles (W, P, X, H, Z, J and L) and one does not (V, S and Q).

Right angle Group (n=8)

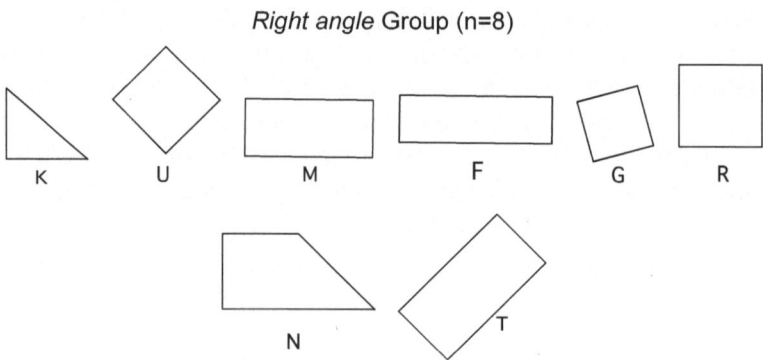

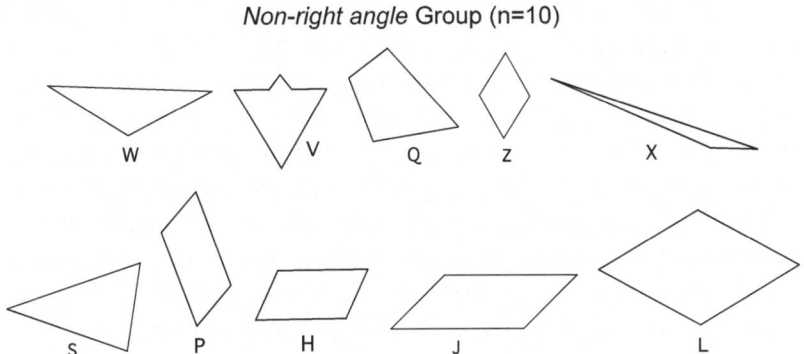

Figure 4.17. Kevin's regrouping of polygons in the pre-interview.

Kevin regrouped the polygons by the attribute of *right angles* or *non-right angles*. The *right angle* group consisted of polygons having at least one right angle, whereas the *non-right angle* group contained all other polygons. Consequently, the *non-rectangle* group contained all other quadrilaterals such as the parallelograms, the rhombi and a trapezoid. To prompt Kevin's understanding of quadrilaterals, the interviewer asked Kevin to subgroup the *non-rectangle* group, at which point he split the group into *rhombus* and *non-rhombus.*

> Kevin: I would separate them into rhombuses. P is a rhombus, L is a rhombus, J is a rhombus, Z and H are all rhombuses. [Pointing at these polygons]

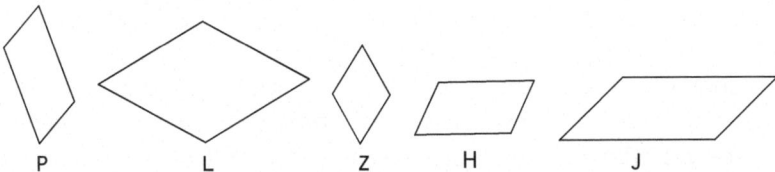

In this group, Kevin identified both parallelograms and rhombi as *rhombus*. When Kevin was asked to identify a parallelogram, he chose H as a parallelogram. When he was asked to identify a rhombus, he pointed to J, another parallelogram in the group. Thus, Kevin revealed his confusion between parallelograms and rhombi.

In the pre-interview, Kevin grouped polygons by the number of sides and by the characteristics of their angles. He also favored dividing polygons into two groups, according to what polygons were and what they were not, with names of polygons such as *rectangle* and *rhombus*. Kevin identified polygons by direct recognition with some assumptions. For example, he mentioned, "just from looking at it... I am assuming that they are right angles" to explain his assumptions. Figure 4.18 summarizes Kevin's *routines of sorting* in the pre-interview.

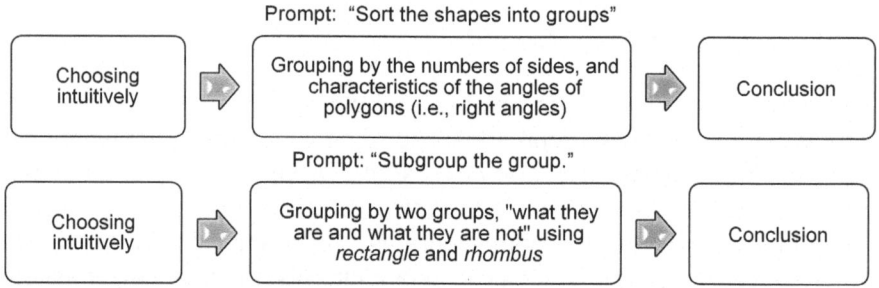

Figure 4.18. Kevin's routines of sorting in the pre-interview.

In the post-interview, Kevin first grouped the eighteen polygons into two groups, consisting of *triangles* (K, W, X and S) and *non-triangles* (the rest of 14 polygons), based on the number of their sides. When Kevin was asked to regroup the eighteen polygons, he regrouped them in two groups: *quadrilaterals* (U, M, F, G, P, N, T, L, J, Z, Q, H and R) and *non-quadrilaterals* (K, W, X, V and S), again by the numbers of sides. The researcher asked him to subgroup the *quadrilaterals*, and he split the *quadrilaterals* into two groups: *parallelograms*, a collection of quadrilaterals with two pairs of parallel sides, and *non-parallelograms*, one trapezoid and one quadrilateral. When Kevin was prompted to subgroup the *parallelograms*, he split the group into *rectangles* and *non-rectangles*. This pattern continued with the two other subgroups consisting of *squares* and *non-squares* within the *rectangles* group. For example, Kevin identified the *non-rectangles* as *parallelograms*, a group consisting of parallelograms and rhombi.

> Researcher: What are the non-rectangles?
> Kevin: The parallelograms? P, H, Z, J, and L. [Pointing at these polygons]

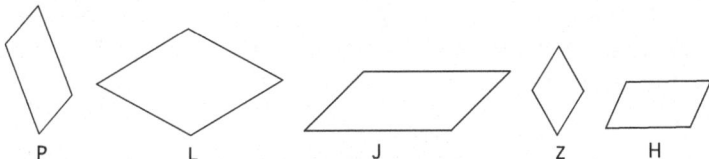

> Researcher: Why they are not rectangles?
> Kevin: Because they don't have four right angles.

Kevin also identified rhombi as a subgroup of *non-rectangles* (i.e., parallelograms), and identified a *square,* a polygon from a group of *rectangles,* as a *rhombus.*

> Researcher: Can you identify a rhombus?
> Kevin: A rhombus? I think Z and L are rhombuses. And then U, G, and R would be rhombuses as well. [Pointing at Fig. Z and L]

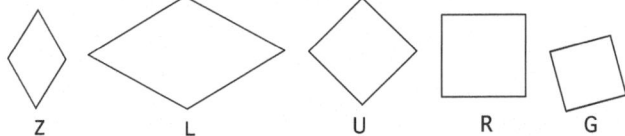

> Researcher: How do you know L is rhombus?
> Kevin: Because these sides look equal in length and they are also parallel to the opposite   sides [pointing at L]
> Researcher: How do you know U is a rhombus?
> Kevin: Because all sides are equal and opposite sides parallel to each other.

In the post-interview, Kevin first grouped the polygons by the numbers of their sides. When asked to regroup the polygons, he classified quadrilaterals by dividing them into two groups each time. During the interview, Kevin did not use measurement tools to verify congruent angles or sides of the polygons; instead he assumed the angles of the rectangles to be 90 degrees, and the sides of the parallelograms to be parallel. It can be concluded that Kevin identified polygons and grouped them intuitively, but it was not self-evident. Table 4.20 illustrates Kevin's grouping of quadrilaterals (non-parallelograms verses parallelogram) during the post-interview, and Table 4.19 summarizes his routines of sorting in the post-interview.

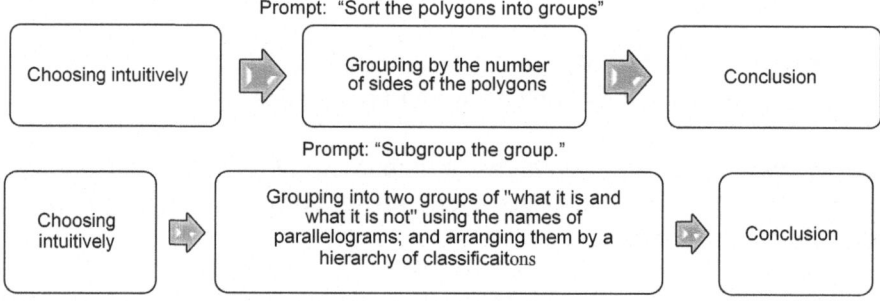

Figure 4.19. Kevin's routines of sorting in the post-interview.

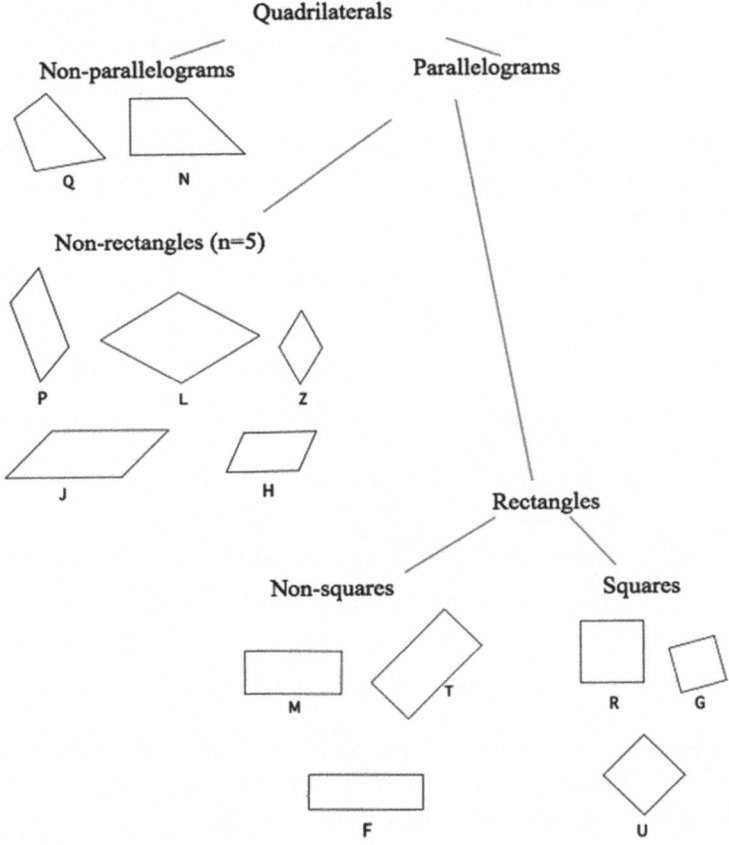

Figure 4.20. An illustration of Kevin's grouping of quadrilaterals in the post-interview.

Kevin had a similar routine pattern when he grouped polygons by the number of sides, and sub-grouped polygons by dividing them into two groups each time during the two interviews. However, in the subgroups, quadrilaterals were arranged in a hierarchy of classifications at the post-interview, in contrast to identifying the *rectangles* and the *rhombi* without a hierarchy in the pre-interview.

Kevin's change in geometric discourse also showed in his routines of substantiation. For example, he changed his routines from *recalling* and *measuring* routines in the pre-interview, to substantiate his claims using endorsed narratives in the post-interview. In the following section, the analyses of changes in Kevin's routines of substantiation are reported.

In the pre-interview, Kevin drew a parallelogram (i.e., a rhombus), and stated that it was a parallelogram because the opposite sides were parallel to each other.

A. Draw a <u>parallelogram</u> in the space below.

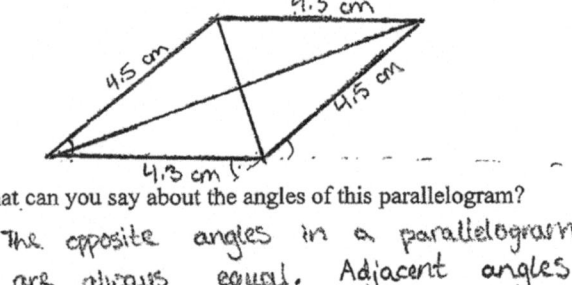

1. What can you say about the angles of this parallelogram?

The opposite angles in a parallelogram are always equal. Adjacent angles are supplementary (sum is 180°).

Figure 4.21. Kevin's drawing of a parallelogram in the pre-interview.

When discussing the angles of the parallelogram, Kevin replied, "I think that is just a property of a parallelogram." The following conversation took place after he was prompted for verification.

> Researcher: Can you show me why opposite angles are equal?
> Kevin: You mean... prove it to you, that in every case it would be that way?
> Researcher: Yes.
> Kevin: I could just measure the angles for you, with a protractor. I've never done a proof before. I've done lots of proofs, but not on something like that [in geometry].

Writing a proof about the angles of a parallelogram was new to Kevin at the time of pre-interview, but he was aware of the difference between a verification of a statement "opposite angles are equal in a parallelogram" in general (i.e., a mathematical proof) and a verification of an example of a statement (i.e., check the measurements of angles), as he responded, "prove it to you that in *every* case it would be that way?" and later proposed to measure the angles.

Kevin also provided another narrative about the angles of a parallelogram that the "adjacent angles are supplementary" (add up to 180°). When asked for substantiation, Kevin replied, "I have learned it before, but I am not a hundred percent sure," and provided his verification.

> Kevin: This angle looks like it would match up if I extend this line out like this [extending the base of the parallelogram]

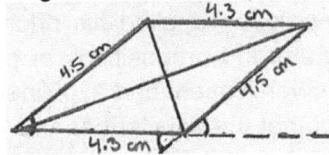

> Kevin: this angle looks equivalent to this angle. And, actually it is, because I've learned about parallel lines [pointing at the corresponding angles]. It would, because this angle would be equivalent to this angle [pointing at the vertical angles] and is also equal to that [pointing at alternating interior angles].

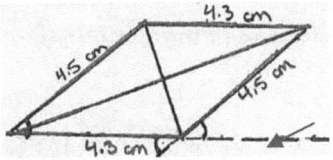

> Researcher: Why is that?
> Kevin: That's another property. I just remember in high school, we learned a lot of properties about parallelograms and parallel lines, and things like that.
> Researcher: So... What is your conclusion?
> Kevin: That the adjacent angles add up to 180 degrees.

Kevin reasoned his way through by drawing extended lines and by identifying parallel sides and congruent angles. He remembered the properties of a parallelogram and the results of parallel line postulate propositions, as well as the relation between the vertical angles, and applied them in his verification process. Kevin identified more congruent angles than he needed for the verification. More specifically, Kevin needed only one of the three congruent pairs of angles, but he identified three pairs: vertical angles, alternating interior angles and corresponding angles created by parallel lines and a transversal. Kevin neglected to point out how the congruent angles would lead to the proof that the adjacent angles add up to 180 degrees, an argument crucial in this substantiation. In this case, Kevin's routine of substantiation of the claim, "the adjacent angles add up to 180 degree" was incomplete.

In the post-interview, Kevin applied his prior knowledge about the properties of parallelogram and propositions of parallel lines to identify the elements needed for verification, and explained why opposite angles are equal, using the fact that the adjacent angles in a parallelogram add up to 180° as an endorsed narrative.

Researcher: How do you know that the adjacent angles add up to 180 degrees?

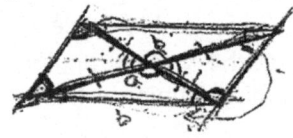

Kevin's drawing at the post-interview

Kevin: Because that's one of the properties of a parallelogram, I can show you.

Researcher: Go ahead.

Kevin: If you were to rip this in half [drawing a line cut through the polygon in half],

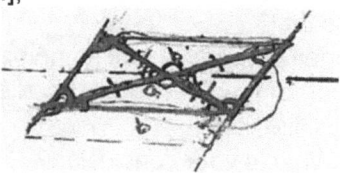

Kevin: and then take this top half and put it down here, then it would line up, like this. Does that make sense?

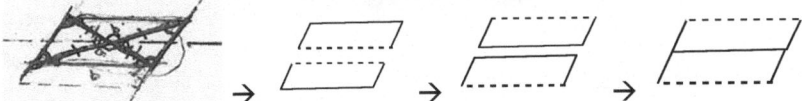

Kevin: Right here… Let's say that this is *a* and this is *b*. If we moved this [b] down here then these would be on the same line. [Labeling the polygons with a and b representing two pieces of the polygon

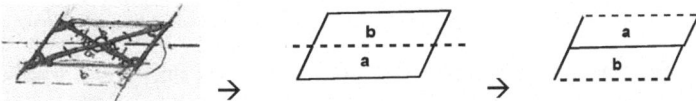

Kevin: This angle here [pointing at the vertex angle] would be right here, and they'd be on the same line. And, angles on a line add up to 180 degrees.

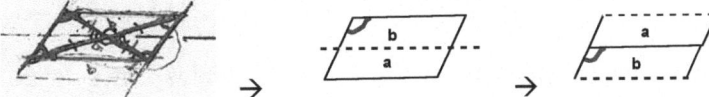

In this excerpt, Kevin first referred to the claim as a property of parallelograms. To verify that adjacent angles add up to 180 degrees, he first divided the polygon horizontally in half by drawing a dashed line to cut the parallelogram in half. Then he described a set of imaginary movements to show that after mappings the adjacent angles would form a straight angle, so they added up to 180 degrees. Later, Kevin used this newly endorsed narrative that adjacent angles add to 180 degrees to justify another claim he made about the angles of the parallelogram: "the opposite angles in a parallelogram are congruent."

Researcher: How do you know that the opposite angles are equal?
Kevin: Because you have two parallel lines, here, if I was to extend these lines [Extending two sides of the parallelogram].

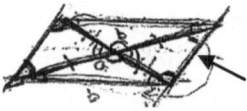

Kevin: This would be a transversal.

Kevin: So, you've got this angle here and this and this angle here add up to 180 degrees because they're adjacent.

Kevin: For the same reason this angle and this angle would also add up to 180 degrees

Kevin: So these [pointing at the opposite angles] have to be the same.

Kevin first identified two parallel lines and a transversal, and concluded that adjacent angles add up to 180 degrees. In order to show opposite angles were equal, Kevin first identified two pair of adjacent angles, with one vertex angle included in the previous ones, so that two pairs of adjacent angles shared one common angle, and he then used the fact that both pairs of adjacent angles added up to 180 degrees. Finally, he concluded that the opposite angles had to be the same by applying a transitive relation (i.e., $a + b = 180$ and $b + c = 180 => a = c$).

Comparing the two scenarios in which Kevin tried to show "the adjacent angles [in parallelogram] add up to 180," Kevin's routine procedures were different. He progressed from referring to prior knowledge to justify his claims (i.e., *recalling routines*), which were incomplete, to using endorsed narratives to substantiate his claims completely. Kevin's routines of substantiation also showed that he moved from identifying more elements than needed for verification without logical order, to choosing the exact number of elements necessary to justify statements logically.

In another case, Kevin went from measuring the sides of parallelograms with rulers to check their measurements for congruency at the pre-interview, to identifying congruent triangles using triangle congruent criterions to verify the congruent parts of parallelograms at the post-interview. Kevin drew a parallelogram (i.e., a rhombus) and stated, "the opposite sides are parallel to each other and are equal in length" in the pre-interview. The following conversation took place after the interviewer asked for substantiation.

> Researcher: Can you show me why they [opposite sides] are equal in length?
> Kevin: In this parallelogram? I can measure it. So this is...4.5 centimeters, this is a little less than 4.5. [Using a ruler to measure one pair of opposite sides]... Right, this looks about 4.3. Yeah, about the same. [Measuring another pair of opposite sides].

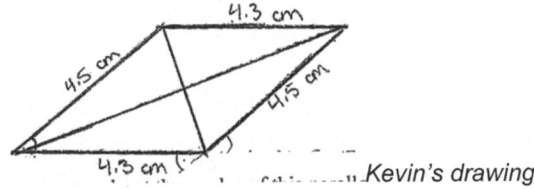
Kevin's drawing

> Researcher: How do you know for *every* parallelogram this is true?
> Kevin: You mean prove it? Well, I am not sure, I know it's just a property of a parallelogram.

To verify the claim, Kevin first referred to it as a property of a parallelogram, and then used a ruler to measure the sides of the parallelogram to check his conclusion. As mentioned earlier, Kevin did not have much experience in constructing a mathematical proof in this context, but was aware of the difference between constructing a proof at an abstract level and checking with a concrete example at an object level. For example, he responded, "In this parallelogram? I can measure it. You mean prove it?"

This pattern of *recalling*, *measuring* and *checking* were consistent during the pre-interview as he was asked to verify the equivalence of diagonals in a rectangle (another parallelogram he drew at the pre-interview.

Researcher: What can you say about the diagonals of this parallelogram?

Kevin: They are of equal length.

Researcher: How do you know that they are equal?

Kevin: Because I learned it a long time ago, in a rectangle, the diagonals are the same.

Researcher: Is there a way that you could convince me?

Kevin: I would measure them, is that O.K? [Using a ruler to measure the diagonals]. Yeah, they're both 8.2 centimeters. So, the diagonals have equal length.

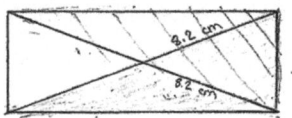

The researcher asked Kevin, "What if you don't have rulers to measure the diagonals, what would you do?" Kevin replied, "If you look, the diagonals form two triangles," and identified two triangles (shaded), and explained why the two triangles were congruent:

Kevin: This side equal to this side [pointing at the opposite sides of the parallleogram]

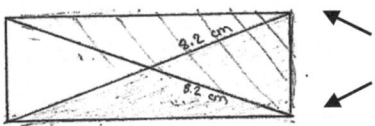

Kevin: This side is obviously it's the same side.

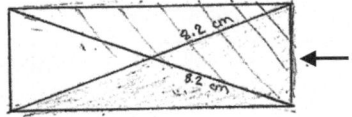

Kevin: Which means this side would have to be equal to this side [pointing at the diagonals]

Kevin intuitively provided a general explanation about the equivalence of the diagonals using congruent triangles: first, he identified two triangles [see shaded areas] where the diagonals were the hypotenuses of the triangles, and then he chose two congruent elements, using the opposite sides of the rectangles as one pair of corresponding sides in the triangles and noting a common side. From there, Kevin concluded that the diagonals were equal based on the equivalence of the two other pairs of corresponding sides of the two triangles. Mathematically, to verify that two triangles are congruent we need three elements, and in this case, Kevin only provided two. The information about an *included* angle of the two sides that Kevin identified is needed to complete the verification. Note that in the earlier conversation, Kevin drew this rectangle as a different parallelogram, and knew that all angles were equal in a rectangle. However, there was no mention of the equivalence of the angles, the third element needed for verification of the congruent triangles.

During the post-interview, Kevin applied a triangle congruent criterion to substantiate the congruent parts of parallelograms. For example, when discussing the diagonals of a parallelogram, Kevin stated, "the diagonals bisect each other." When asked for verification, Kevin provided the following explanations using the Angle-Side-Angle

(ASA) triangle congruency criterion. Table 4.15 illustrates a step-by-step routine from Kevin's explanation with corresponding transcripts.

Table 4.15. Kevin's Routine of Substantiation of Diagonals Bisect Each Other in the Post-Interview

Routine Procedures	Transcripts
1. Identify triangles formed by diagonals	I drew two diagonal and … there are four triangles here.
2. Verification of two congruent triangles 2.1 Identify first pair of corresponding angles of the triangles	If you take this angle here and this angle here, they're equal to each other because they're vertical angles.
2.2 Identify the second pair of corresponding angles of the triangles	Because these two lines are parallel, this angle would be equal to this angle here, they're opposite interior angles, …
2.3 Identify the third pair of corresponding angles of the triangles	The same for this angle and this angles... same property.
2.4 Identify the fourth pair of corresponding angles of the triangles	…and then this angle and this angle.

2.5 Identify one pair of corresponding sides of the triangles	These two sides are equal because that's a property of a parallelogram [pointing at the sides of the triangles] 
4. Verify congruent triangles using A-S-A correspondence	So, you have Angle-Side-Angle here. And, so, that shows that this triangle here is congruent to this triangle here  
5. Conclusion	When you match up the corresponding sides, this side would be congruent to this side and then, this would be congruent to this. So, they are equal measure, so they bisect each other. 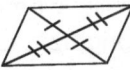

Table 4.15 shows that Kevin used the ASA triangle criterion to verify that two triangles were congruent in his explanations, and concluded that the diagonals bisected each other because the diagonals intersected and created two congruent triangles. Note that at each step, Kevin was able to provide mathematical justifications for his conclusions. For example, Kevin stated, "*because* these two lines are parallel" and "*because* they are vertical angles" as justifications to demonstrate that two pairs of angles were equal. He also used "a property of parallelogram" to explain the congruent sides.

In the process of verification, Kevin identified four pairs of congruent angles (i.e., vertical angle, alternating interior angles) and one pair of congruent sides, more elements than needed for verifying two triangles (shaded areas) were congruent. In this case, Kevin did not need the pair of vertical angles and a pair of alternating interior angles for his verification of two congruent triangles.

As the conversations continued, Kevin drew a rectangle and stated that the diagonals also bisected each other. The researcher asked him for substantiation.

Researcher: How do you know diagonals bisect each other in this case?

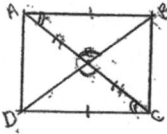

Kevin's drawing

Kevin: For the same reason as last time, do you want me to explain again?

Researcher: When you say, "for the same reason as last time," what do you mean?

Kevin: Just, all of it, when you create these triangles and the triangles are congruent to each other...based on that property, all parallelograms have diagonals that bisect each other [pointing   at the triangles in the parallelogram].

Kevin's drawing

Kevin referred to his previous explanations about why the diagonals bisecting each other in a parallelogram to justify his claim about the diagonals bisecting each other in a rectangle. He connected the pattern—"the diagonals bisect each other"—in both the rectangles and the parallelogram. When the interviewer asked for a written proof, Kevin provided the following:

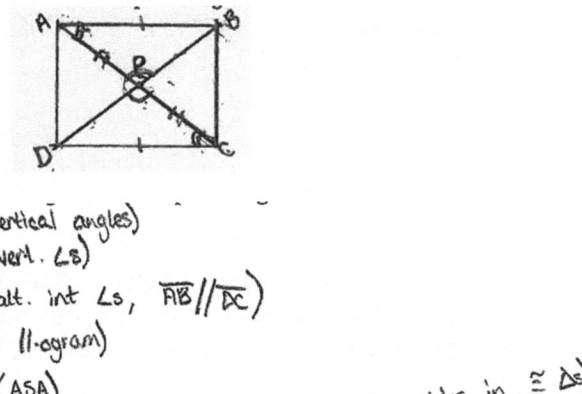

$\angle APB \cong \angle DPC$ (vertical angles)
$\angle APD \cong \angle BPC$ (vert. $\angle s$)
$\angle BAC \cong \angle DCP$ (alt. int $\angle s$, $\overline{AB} // \overline{DC}$)
$AB = DC$ (prop. of ||-ogram)
$\triangle ABP \cong \triangle CDP$ (ASA)
$AP = PC$ (corr. sides in $\cong \triangle s$) $DP = BP$ (corr. sides in $\cong \triangle s$)

p is the intersection of the diagonals

Figure 4.22. Kevin's written proof that diagonals bisect each other in the post-interview.

Before Kevin started to write the proof, he labeled the vertices of the rectangle with A, B, C, and D in a clockwise order. He used mathematical symbols like $\angle$, $\cong$, $\triangle$, $//$ to replace the words *angle*, *congruent*, *triangle*, and *parallel*, respectively, in his written proof. He wrote "$\angle APB \cong \angle DPC$," "AB = DC" and "$\triangle ABP \cong \triangle CDP$" (see Figure 4.22) to indicate two congruent angles, sides and triangles accordingly. The response Kevin gave when asked for further clarification follows.

Kevin: I have $\angle$ APB, so this angle right here, is equal to $\angle DPC$, because they are vertical angles. Then also $\angle APD$, so this here well I guess that's not really important.

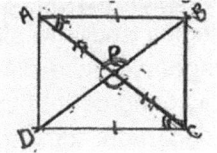 $\angle APD \cong \angle BPC$ (vert. $\angle s$)
$\angle APB \cong \angle DPC$ (vertical angles)

Kevin: ∠BAC is congruent to ∠DCP because AB and DC are parallel to each other, so these two angles are alternate interior angles, and they're always congruent.

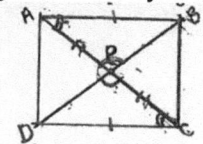

∠BAC ≅ ∠DCP
(alt. int ∠s, AB//DC)

Kevin: AB is equal to DC because that it's a property of a parallelogram. So, we have two angles on this side, and that is enough information to conclude that △ABP is congruent to △CDP.

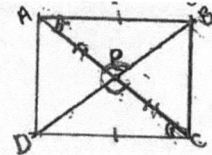

AB = DC (prop. Of parallelogram)
△ABP ≅ △CDP (ASA)

Researcher: When you say two angles and a side, what do you mean?

Kevin: This angle and this angle. And then we have a side.

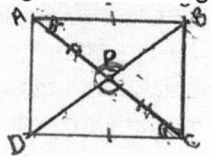

∠APB ≅ ∠DPC
(vertical angles) – *angle*
∠BAC ≅ ∠DCP
(alt. int ∠s, AB//DC) – *angle*
AB = DC (prop. Of parallelogram) – *side*

Kevin's routines of substantiation changed. At the post-interview, he constructed a mathematical proof using mathematical symbols and justifications. In particular, Kevin identified two angles and one side to verify two congruent triangles created by the diagonals. In a different task, Kevin applied ASA congruent criterion to verify congruent triangles. This shows that Kevin did not rely on one congruent criterion to substantiate congruent triangles. It appeared that Kevin was at the beginning stage of constructing formal mathematical proofs with abstraction. More evidence of the change in Kevin's geometric discourse was in his word use.

Changes in Kevin's Word Use
Kevin's use of the words changed from indicating both rhombi and parallelograms as *parallelograms* in the pre-interview, to representing a hierarchy of classifications of parallelograms in the post-interview. Table 4.16 shows the frequencies in which the words *quadrilateral*, *parallelogram, rectangle, square, rhombus, trapezoid,* and *kite* were used in both interviews.

Table 4.16. Frequencies of Kevin's Use of the Names of Quadrilaterals

Name	Frequency				Total Frequency	
	Pre-T1	Post-T1	Pre-T2	Post-T2	Pre	Post
Quadrilateral	0	1	0	4	0	5
Parallelogram	3	4	6	14	9	18
Rectangle	9	3	4	4	13	7
Square	1	3	3	8	4	11
Rhombus	11	3	0	3	11	6
Trapezoid	2	2	0	0	2	2
Kite	0	0	0	0	0	0

As shown in Table 4.16 the word *parallelogram* (n=27) was most frequently used during the interviews. The word *rectangle* (n=21) was the second most frequently used, and *rhombus* (n=17) the third. Kevin did not mention the word *kite* (n=0) at all in the interviews, and *trapezoid* (n=4) was the second least mentioned. There was an increase in use of the words *quadrilateral, parallelogram* and *square* from the pre-interview to post-interview. In particular, the word *quadrilateral* was used only during the post-interview. The word *parallelogram* almost doubled in the post-interview, while the word *square* almost tripled. There was a reduction in the use of the words *rectangle* and *rhombus* at the post-interview. In the following section, the detail of Kevin's use of the word *parallelogram* at the interviews is reported.

Kevin's routines of sorting polygons also revealed his understanding of the concepts of the quadrilaterals. In the pre-interview, he identified all 4-sided polygons with opposite sides that are parallel and opposite angles that are equal, as *rhombi*.

Kevin: P is a rhombus, L is a rhombus, J is a rhombus, Z and H are all rhombuses.

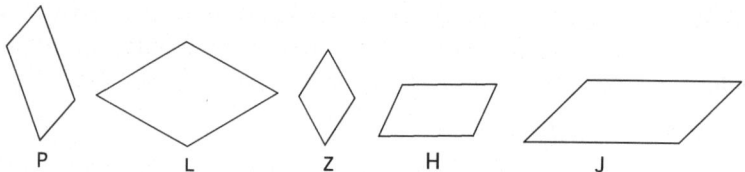

Researcher: What is a rhombus? [Pointing at J]

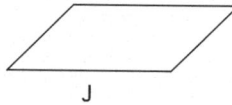

Kevin: A rhombus is a four-sided figure, and the opposite sides and angles are equal.

Researcher: What is a parallelogram?

Kevin: A parallelogram is four-sided figure with opposite sides and angles are equal. This is a parallelogram [pointing at Fig. J].

Kevin's definitions of rhombus and parallelogram showed no clear distinction between the two, as he described both as "a four-sided figure[s] with opposite sides [that] are parallel and opposite angles [that] are equal." When the researcher asked Kevin to draw a parallelogram, he provided a drawing with the properties of a rhombus (a quadrilaterals with all four sides have the same measure).

Researcher: Why do you think this is a parallelogram?

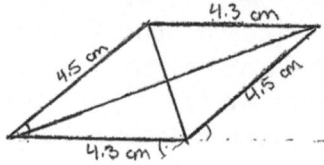

Kevin: Because opposite sides are equal and parallel.

Researcher: What can you say about the diagonals of this parallelogram?

Kevin: They are perpendicular.

Kevin drew a rectangle as a different parallelogram in the pre-interview. He defined a rectangle as "a four-sided figure with all four angles are 90° and opposite sides are parallel and equal in length." In the pre-interview, Kevin showed a connection between parallelograms, rectangles, and rhombuses. Recall that Kevin grouped squares and rectangles together as *rectangles*, and defined square as "a four-sided figure with all angles 90° and all sides are equal." It can be concluded that Kevin's use of the word *parallelogram* applied to *rhombuses/ parallelograms,* a family of four-sided figures that have opposite sides equal and parallel and *rectangles*, a family of four-sided figures that have all 90° with opposite sides equal and parallel. Although Kevin included squares as rectangles, he did not separate squares from rectangles as an individual group. Figure 4.23 illustrates Kevin's use of the word *parallelogram* in the pre-interview.

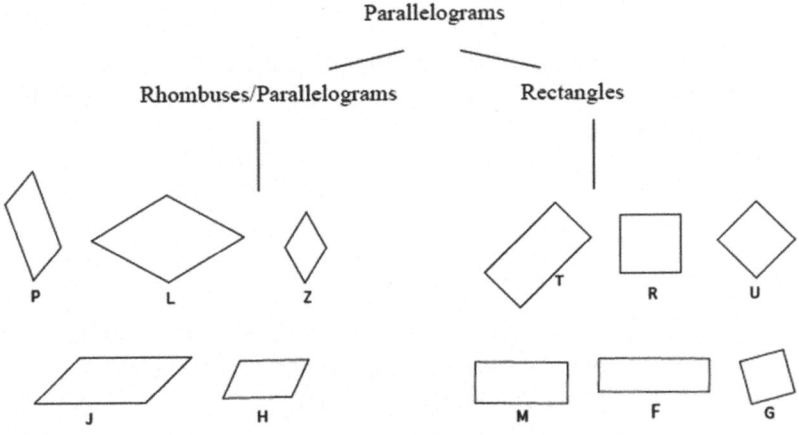

Figure 4.23. Kevin's use of the word parallelogram in the pre-interview.

In the pre-interview, Kevin's use of the words *parallelogram* referred to a family of four-sided polygons that share a common descriptive narrative: opposite sides are equal and parallel. Kevin did consider squares as rectangles, but he made no connections between a rhombus and a square. His grouping and identification of quadrilaterals suggests that he knew the definitions of each parallelogram and applied them, but did not show an understanding of a hierarchy of classifications of quadrilaterals.

In contrast, Kevin demonstrated the understanding of the word *parallelogram*, and revealed it in the hierarchy of the classifications of parallelograms at the post-interview. For example, Kevin included all rectangles, squares, rhombuses, and parallelograms as *parallelograms* among the eighteen polygons. He also split the *rectangles* group to rectangles and squares.

> Researcher: Can you show me what are the rectangles?
> Kevin: Assuming that these are 90° angles. U, M, F, G, T, and R are rectangles.

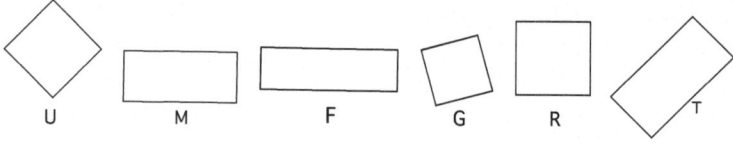

> Researcher: Can you subgroup the rectangles?
> Kevin: Into squares and non-squares. So the squares would be U, G, and R.

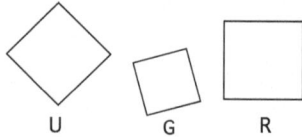

When Kevin was prompted identify a rhombus and a square, he also made connections between rhombuses and squares:
Researcher: How do you know L is a rhombus?

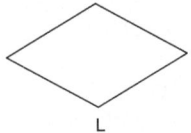

L

Kevin: All sides are equal in length and opposite sides are parallel to each other.
Researcher: Is R a rhombus?
Kevin: Yes.
Researcher: Why?
Kevin: Because all sides are equal and opposite side are parallel to each other.

During the interview, the researcher asked Kevin to write down the definitions of parallelograms. Kevin's definitions of parallelograms are provided in Table 4.17.

Table 4.17. Kevin's Definitions of Parallelograms in the Post-Interview

Name	Definitions	Sample Polygons
Square	Four-sided figure in which all four sides are equal length and all angles are right angles.	U R
Rectangle	Four-sided figure in which all four angles are right angles.	M G
Parallelogram	Four-sided figure in which both pairs of opposite sides are parallel to each other.	H M

Rhombus	Four-sided figure in which all sides are equal and opposite sides are parallel	◇ □ Z G

In the post-interview, Kevin used the word *parallelogram* to describe a collection of quadrilaterals with different appearances and names, and arranged by the characteristics of their angles (i.e., right angle versus non-right angle) and sides (i.e., all sides equal versus opposite sides equal). Although given different names such as *rectangles, parallelograms, rhombi,* and *squares,* they are all called *parallelograms* because they fit the description of opposite sides being equal and parallel. Figure 4.24 illustrates Kevin's understanding of the concept of quadrilaterals with a hierarchy.

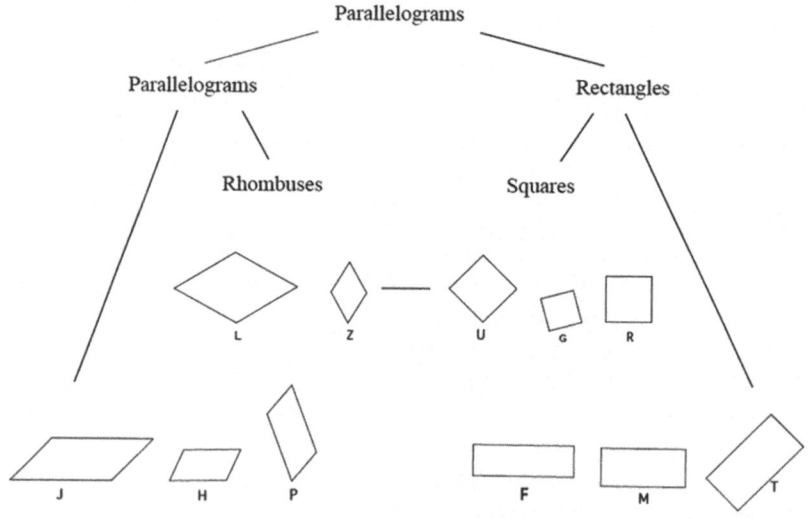

Figure 4.24. Kevin's use of the word parallelogram in the post-interview.

As shown in Figure 4.24, Kevin identified quadrilaterals which was determined by "what it was and what it was not" proceeding through families of quadrilaterals such as *parallelograms, rectangles* and *squares.* During the interview, Kevin demonstrated his understandings of the relations between *parallelograms, rectangles, rhombuses,* and *squares* by classifying these quadrilaterals in a hierarchy.

The results suggest that Kevin's geometric discourse changed from relying on the measurements of angles and sides to verify their congruency, to using axioms and propositions to substantiate claims about the congruent parts of the parallelograms. Although Kevin's mathematical proofs were not all correct and sometimes were incomplete, he demonstrated an ability to construct mathematical proofs using symbols and justifications. Kevin's use of names of parallelograms also changed to a structured hierarchy of classifications of parallelograms in the post-interview.

Among the twenty pre-service teachers who participated in the pre- and post-interviews, five of them showed *no change* in their van Hiele levels according to the van Hiele Geometry pre-test and post-test responses. Among the five pre-service teachers, one stayed at van Hiele level 2 (Sam), and four stayed at van Hiele level 3 (Judi, Chris, Mandy and Kathy). However, the interview analyses showed the changes in their geometric discourses. To describe these changes, findings of Sam and Judi's geometric discourses are reported in the following sections.

4.2.4 Case Four: Changes in Sam's Geometric Discourse

Sam was a second year college student at the time of the study. Sam took her last geometry class seven years prior to the interview. The van Hiele Geometry pre- and post-tests placed her at Level 2, a descriptive level where polygons are recognized by their properties without orders. A summary of findings about the changes in Sam's geometric discourse is as follows:

- Sam's routines of substantiation were verbal descriptions about the processes of activities using transformations such as reflections, translations, and rotations at the pre-interview, and involved constructing mathematical proofs using propositions and definitions at the post-interview.

- Sam attempted congruent criterion such as Side-Side-Angle (SSA) which was invalid to verify congruent triangles at the pre-interview, and used multiple and valid congruent criterions to substantiate congruent triangles at the post-interview.
- Sam's use of the word parallelogram signified all polygons have pairs of parallel sides without considering all the necessary conditions at the pre-interview, and identified parallelograms as quadrilaterals who have two pairs of parallel sides without a hierarchy of classification.

Changes in Sam's Routines

During the pre-interview, while Sam was sorting the polygons into groups, she asked, "Am I doing it on the assumption that those are right angles [pointing at the angles of a square], and can I assume anyway?" Sam's first attempt at sorting geometric shapes was based on "the numbers of sides they had" and she called them: 1) 3-sided figures (K, W, X, and S), all triangles; 2) 4-sided figures, all quadrilaterals; and 3) 6-sided figure (V), a hexagon. When Sam was prompted to subgroup the 4-sided group, her first reaction was, "If I can assume that the sides appear to be parallel to each other," while pointing to the opposite sides of a parallelogram. Sam arranged the 4-sided shapes into three subgroups and called them Group 1, squares and rectangles; Group 2, parallelograms; and Group 3, a quadrilateral and a trapezoid. Figure 4.25 details the three subgroups of the quadrilaterals.

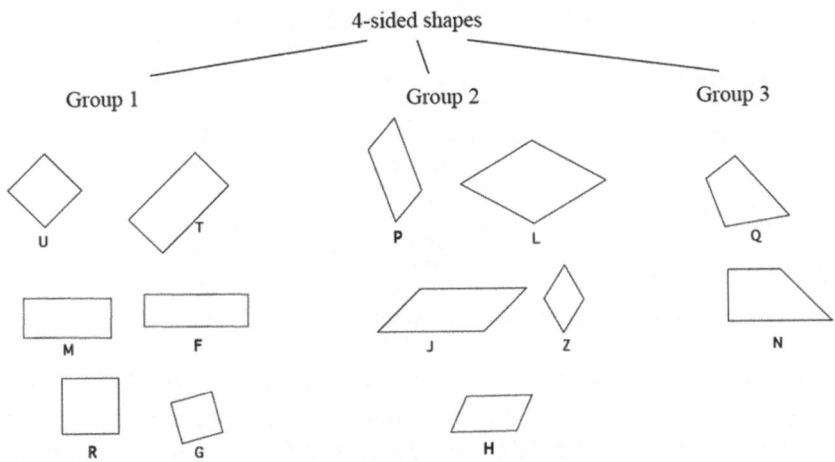

Figure 4.25. Sam's subgrouping of quadrilaterals in the pre-interview.

As an illustration, Figure 4.25 shows the three subgroups for quadrilaterals Sam provided: squares and rectangles (Group 1), parallelograms (Group 2), and a group of 4-sided figures that do not fit into the descriptions of the two previous groups (Group 3). Sam also made clear about the characteristics of each group. For example, she talked the *parallelograms group* consisting only of the parallelograms that "don't have right angles," and the *squares/rectangles group* consisting of figures that "have four sides, all right angles, pairs of sides are parallel and have the same length." Sam's routines of sorting focused on the attributes of angles (e.g., right angles) and sides (e.g., parallel sides or equal sides). During the interview, Sam did not use tools to check the measurements of the angles and sides of the figures instead all her claims about the figures fell under the assumptions of "the sides appeared parallel" and "angles are right angles." Figure 4.26 illustrates Sam's routines of sorting polygons into different groups at the pre-interview.

First prompt: "Sort the shapes into groups"

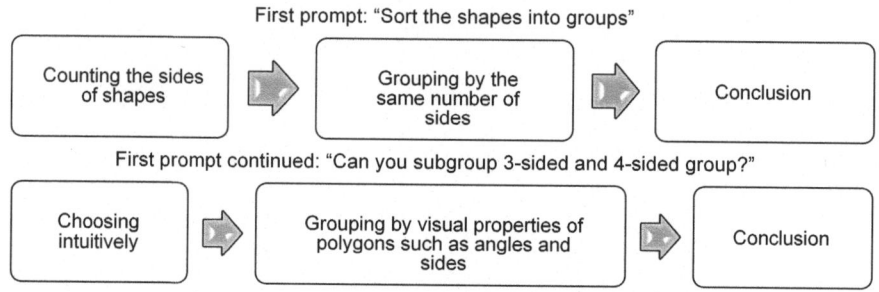

First prompt continued: "Can you subgroup 3-sided and 4-sided group?"

Figure 4.26. Sam's first prompt to group polygons in the pre-interview.

When Sam was asked to regroup the figures differently, her first response was, "I want to separate them into shapes containing right angles and shapes that do not contain right angles." Among the eighteen polygons, Sam split them into two groups: polygons with at least one right angle (Group1), and polygons with no right angles (Group 2). For this grouping, Group 1 is a collection of squares, rectangles a right trapezoid and a right triangle; and Group 2 includes triangles, quadrilateral, parallelograms, and a hexagon (see Figure 4.27).

Group 1: polygons have at least one right angle

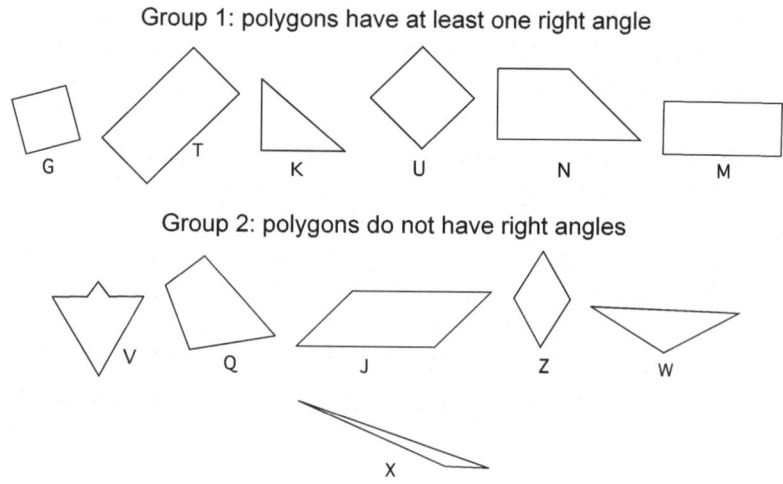

Figure 4.27. Examples of Sam's regrouping in the pre-interview.

Sam divided polygons that have right angles from those figures that do not have right angles. The researcher asked her for subgrouping again.

Researcher: Can you subgroup Group 1?
Sam: I guess for Group 1 [subgroup 1], I could take squares and non-squares; I could take shapes that have acute angles like K and N.

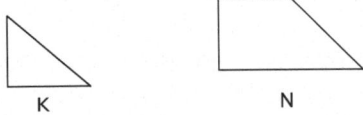

Sam: I am sure that by defining the second group [subgroup 2]. I don't know they kind seem exclusive... will follow figures that don't contain acute angles [U and M].

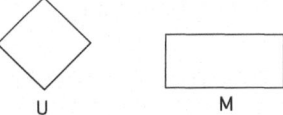

She then divided Group 2 (the polygons do not have right angles) into two subgroups that: Subgroup 1 with polygons that have at least one set of parallel sides, and subgroup 2 with polygons that have no parallel sides. Figure 4.28 illustrates the two subgroups in Group 2.

Subgroup 1: with at least one set of parallel sides

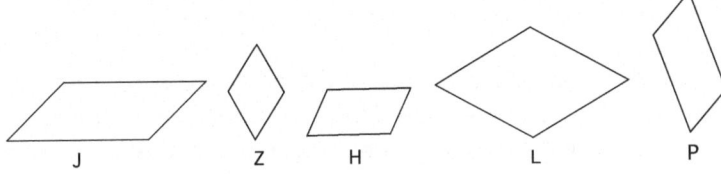

Subgroup 2: with no parallel sides

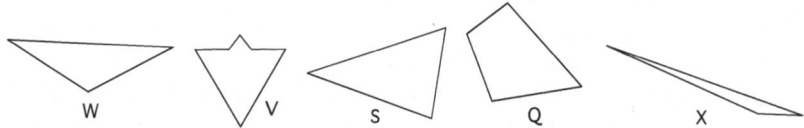

Figure 4.28. Two subgroups of Group 2 in the pre-interview.

Sam's routines of sorting geometric shapes focused on the attributes of the angles of polygons, *not* the sides. She first divided all polygons (n=18) into two groups by the attributes of right angles; and then divided Group 1 into two subgroups based on the property of acute angles. Sam divided Group 2 by the attributes of parallel sides provided half of the polygons in Group 2 are parallelograms. Figure 4.29 summarizes Sam's routine procedures at the second attempt.

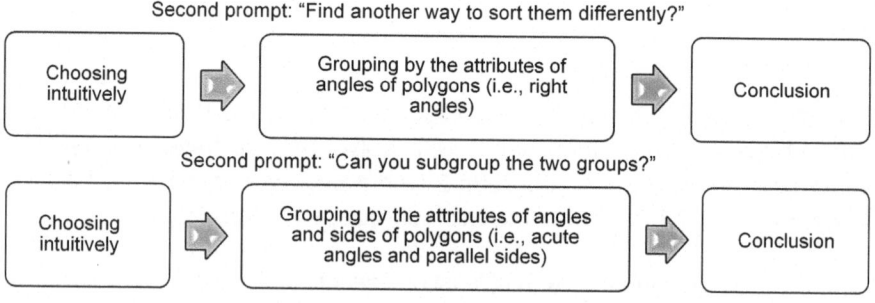

Figure 4.29. Sam's second prompt to regroup polygons in the pre-interview.

Sam's routines of sorting polygons at the post-interview were similar to what she did in the pre-interview. However, there were changes in her routines of substantiation and the analyses of these changes are provided in the following section.

In the pre-interview, Sam drew a parallelogram and stated, "all the angles should add up to equal 360°" (see figure 4.30).

A. Draw a **parallelogram** in the space below.

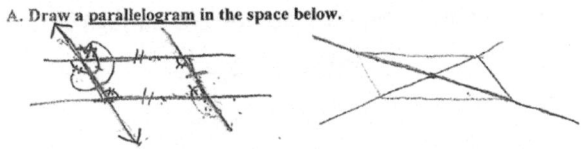

1. What can you say about the angles of this parallelogram?
They all All the angles should add up to equal 360°

Figure 4.30. Sam's drawing of a parallelogram in the pre-interview.

The researcher prompted for verification by asking, "How do you know that all angles add up to 360°?" Sam responded the following:

> Sam: When you have parallel sides, you can extend all the sides…
> [She extended the sides of the parallelogram]

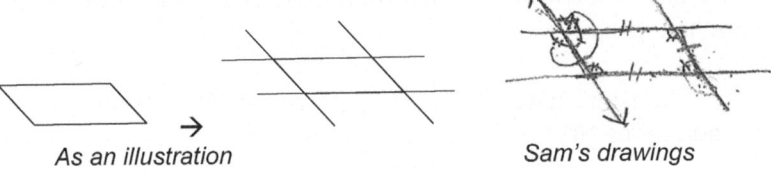

As an illustration   *Sam's drawings*

> Sam: It's 180° and they're complementary angles. [Pointing at the two angles that form straight line].

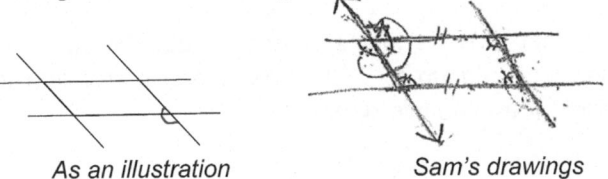

As an illustration   *Sam's drawings*

> Sam: but you can see that this angle really just match this angle

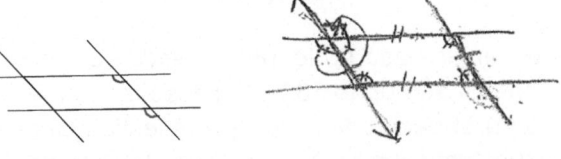

As an illustration   *Sam's drawings*

Sam: So you know that these two angles together are equal 180°.

As an illustration   *Sam's drawings*

In the preceding excerpts, Sam first drew extended lines on the sides of the parallelogram, by saying "you have parallel sides... you can extend..." and identified a vertex angle and its corresponding exterior angle forming a "straight angle." She then identified an adjacent vertex angle transversal to the same exterior angle and made an intuitive claim about the two angles: "you ...see this angle...matches this angle." Sam concluded that the two adjacent vertices of a parallelogram added up to 180°. She also used the endorsed narrative: "two angles equal 180°" to finish the final claim:

Sam: From this diagram and the parallel sides, these two angles add up to 180°

Sam: The fact that it's just a mirror image, the two sets of 180° angles just add up to 360°. [Marking a mirror line to split the parallelogram into two parts]

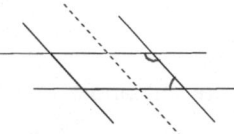

Sam used "two angles add up to 180" to endorse a new narrative, "two sets of 180° angles add up to 360°" because these are a "mirror image" (i.e., a reflection) of each other, using a reflection line (i.e., the dashed line). Sam's substantiation of the narrative, "all angles add up to 360°" was intuitive and self-evident because the reflection line that Sam constructed was not a line of reflection of the parallelogram. Mathematically, this parallelogram only has rotational symmetry, with respect to the center of the parallelogram (i.e., where the diagonals intersect), and not line symmetry. Sam used a "mirror image" (i.e., reflection) to draw a conclusion of all the angles add up to 360° was invalid.

During the pre-interview, Sam frequently used reflections, rotations, and translations to substantiate her claims. For example, she was asked to verify her claim, "two opposite angles (i.e., $\angle$ 1 and $\angle$ 4) are equal."

Researcher: How do you know this angle is equal to this? [Pointing at $\angle$1 and $\angle$4].

As an illustration   *Sam's drawings*

Sam: This angle [$\angle$1] can just be slid over to this position and create this angle [$\angle$2].

Sam: This line [drawing arrowhead on the line] can be rotated so that this angle [$\angle$2] now becomes this angle [$\angle$3].

Sam: This angle [$\angle$3] at this intersection, can just be slid down and then be in this angle's position [$\angle$4].

Sam: So these two angles are equal [$\angle$1 and $\angle$4].

In this case, Sam used words such as "slid over," "rotated" and "slid down" to indicate a sequence of transformations performed on the angles and sides of the parallelogram to substantiate the claim that "two opposite angles are equivalent." Lines and angles are static mathematical objects, but Sam used sequences of imaginary movements to complete her substantiation; through Sam's descriptions, the imaginary movements became visible to others. Sam had an active view of static geometric objects (i.e., angles, sides, etc.) in her substantiation and provided informal proofs through the use of transformations in the pre-interview.

In the post-interview, Sam used mathematical propositions and axioms to verify her claim of "all angles measure [in a parallelogram] add up to 360°."

> Sam: Angles on a straight line add up to 180° [Extending one side of the parallelogram with a dashed line, and pointing at the two angles].

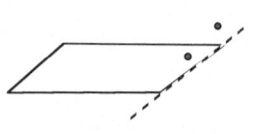 

As an illustration   *Sam's drawings*

> Sam: This angle here is the same as this angle because parallel lines meet a third line at the same angle.

As an illustration   *Sam's drawings*

> Sam: By the same reason, this angle added to this angle equals 180.

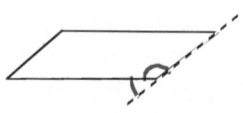 

As an illustration   *Sam's drawings*

> Sam: These two also add up to 180°.

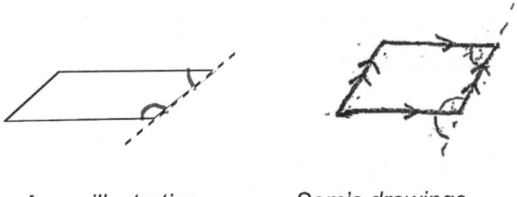

As an illustration   Sam's drawings

Sam: for a similar reason, these two angles add up to 180°

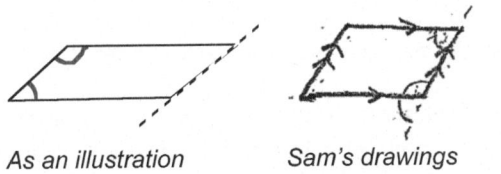

As an illustration   Sam's drawings

Sam: Together they [all four angles in a parallelogram add up] equal 360°.

In contrast to Sam's routines of substantiation at the pre-interview, this example shows two changes that are evident. The first is that in each step of substantiation, Sam provided endorsed narratives (e.g., mathematical axioms and propositions, etc.) as evidence instead of reasoning intuitively. For example, Sam explained how two transversal angles are equivalent, not because you "can see it" (in the pre-interview), but as a result of "two parallel lines meet a third line at the same angle." The second change occurs in Sam's conclusion that "all angles add up to 360°." In the pre-interview, she used "mirror image" which was invalid to verify the claim, whereas at the post-interview Sam reached her conclusion in repeating of similar proofs that "two angles add up to 180°" for two adjacent angles in a parallelogram. Thus, Sam's routine of substantiation shifted to constructing formal proofs using endorsed narratives. This maturity is also revealed in Sam's substantiation of congruent triangles. In the following example, the changes in Sam's routines of substantiation of two congruent triangles are provided.

In the pre-interview, Sam described the diagonals of the parallelogram, saying "the diagonals intersect in the center of the figure,

and divide each other into two equal halves at the intersection point."
She provided the following diagram:

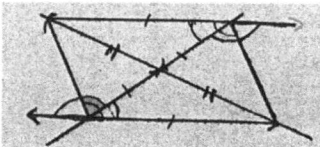

Figure 4.31. Sam's drawing of diagonals intersecting at the center of the parallelogram.

When asked for substantiation, Sam talked about diagonals creating two
pairs of congruent triangles. She identified one pair of such congruent
triangles, and then identified two corresponding sides and two
corresponding angles from the two triangles to verify their congruency.

> Sam: It is given that these are parallel sides. And, these angles are
> equal [adding two angle signs].

As an illustration

> Sam: It's essentially the same intersection, translated to a new
> position [extending lines].

As an illustration

> Sam: This angle is the same as this angle [adding arrowheads on
> the two extended lines]

As an illustration

> Sam: And the complementary angles, the smaller angle that makes
> it add up to 180° [identified two angles that form a striaght angle]

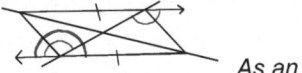
As an illustration

Sam: Is the same over here [identifying another two angles that form a straight angle].

As an illustration

Sam: So, now I know that the angle here of this triangle is equivalent to the angle here of this triangle [pointing at the alternating interior angles].

As an illustration

Sam: And this side length is, the same of this side length. So, I've already shown how a side length and an angle match of each [referring to the two sides].

As an illustration

Sam: And then diagonals bisect themselves equally. I can't really prove that, but I know this side length is the same as this side length [adding two marks on the diagonal].

As an illustration

Sam: this triangle is equivalent to this triangle here [The shaded area indicates two congruent triangles].

As an illustration

Sam's substantiation of two congruent triangles included two parts: the first was the substantiation of the equivalence of alternating interior angles, and the second was the verification of congruent triangles. In the first part of her substantiation, "this angle is equivalent to this angle" (i.e., alternating interior angles). Sam implied opposite angles were equivalent in a parallelogram. She used an instinctive process of translating the intersection to a new position, and then suggested that the corresponding alternating exterior angles were equivalent as a result of straight angle theorem. The second part of substantiation involved the verification of the two congruent triangles (shaded in the diagram) she identified. She identified three elements needed to verify congruent triangles. However, Sam's choice of three elements (angle, side, side) for verification of congruent triangles was invalid, because the choice of angle-side-side does not always guarantee congruent triangles.

In the post-interview, Sam was able to use triangle congruency criterion to substantiate most of her claims "opposite angles are equivalent," "opposite sides are equivalent" and "diagonals bisect each other" in a parallelogram. Sam applied Side-Angle-Side and Angle-Side-Angle, to verify congruent triangles, and she was comfortable using these methods. The following response illustrates Sam's substantiation of "diagonals bisect each other."

> Sam: I'm looking at this triangle as compared to this one here [shaded triangles]

As an illustration

> Sam: And I know that these two angles are congruent [two marked angles] because between these parallel lines, and now this diagonal [two parallel sides and a diagonals a transversal].

As an illustration

> Sam: These angles are also congruent [two other angles].

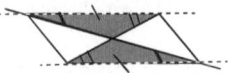

As an illustration

Sam: By the triangle test, angle, side, angle, these two triangles [shaded] are congruent.

Sam: which means that this side corresponds with this side and that this side corresponds with that side [each half of the diagonals].

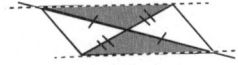 *As an illustration*

In the preceding substantiation Sam first verified that the "two triangles are congruent" using the angle-side-angle criterion by identifying two corresponding angles and their included sides. She used the endorsed narrative "the two triangles are congruent" to construct a new narrative that "diagonals bisect each other," by saying "this side corresponds with this side" as a result of congruent triangles. Sam was familiar with the alternating interior angles theorem, as she explained:

Sam: We know that between parallel lines, if you take a third line and cross both lines, then it will have [alternating interior] angles that are congruent. In this case, this angle and this angle.

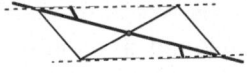 *As an illustration*

During the post-interview, Sam applied the same substantiation to other similar situations. For instance, Sam applied the same reasoning to show diagonals bisect each other in a rectangle, as she responded, "the same as what I did in parallelogram, I already established that." When she was asked "is it true that in *all* parallelograms diagonals bisect each other?" She responded, "Yes, that's true" and shared her thinking:

Sam: because when you draw the diagonals in a figure [parallelogram], there is an intersection point and it divides the figure into four triangles. Regardless of the figure, if it's a parallelogram, these two triangles will be congruent and these two triangles [two pairs of opposite triangles created by the diagonals] will be congruent. So, it can be found that in congruent triangles, corresponding sides will be equal [therefore diagonals bisect each other in all these cases].

The changes in Sam's substantiation routines showed that she gained more familiarity of constructing formal proofs using mathematical axioms. She was able to generalize characteristics among parallelograms, such as diagonals bisect each other in all parallelograms. Sam also showed changes in her use of mathematical words.

Changes in Sam's Word Use

Sam's use of the word parallelograms changed as she gained more familiarity with the quadrilaterals and their attributes over the ten weeks period. She used the words *quadrilateral, parallelogram, rectangle, square, rhombus, trapezoid,* and *kite,* in both interviews (see Table 4.18). As shown in Table 4.18, the word *parallelogram* (n=64) is the most frequently used during both interviews. The word *square* (n=22) is the second most frequently used, and the word *rectangle* (n=8) is third. Note the large difference in the frequency of the words rectangle and *parallelogram* (n=56), and between the words *parallelogram* and *square* (n=42). The word *kite* was not mentioned at all in both interviews. The finding shows the word *rhombus* and *trapezoid* were only mentioned at the pre-interview. There was a reduction in use of the words *quadrilateral, parallelogram, rhombus,* and *trapezoid* at the post-interview, and there was a slight increase in use of the word *rectangle* at the post-interview. However, the frequencies of the word counts do not provide details about how these words were used. The following analysis looks at the changes in Sam's word meaning in the use of *parallelogram, rectangle, square, trapezoid* and *rhombus.*

Table 4.18. Frequencies of Sam's Use of the Names of Quadrilaterals in the Two
            Interviews

Name	Frequency				Total Frequency	
	Pre-T1	Post-T1	Pre-T2	Post-T2	Pre	Post
Quadrilateral	3	1	0	0	3	1
Parallelogram	7	5	33	19	40	24
Rectangle	3	4	0	1	3	5
Square	5	4	6	7	11	11
Rhombus	6	0	0	0	6	0
Trapezoid	3	0	0	0	3	0
Kite	0	0	0	0	0	0

During the pre-interview, Sam grouped eighteen polygons based on the attributes of right angles and parallel sides. For example, she split the polygons into two groups, one containing at least one right angle, and one that did not. She also split the polygons into a group of polygons with parallel sides and the one group of polygons with no parallel sides. Sam's misunderstanding of parallelograms was detected when she started to draw different parallelograms.

Researcher: Why is this a parallelogram?

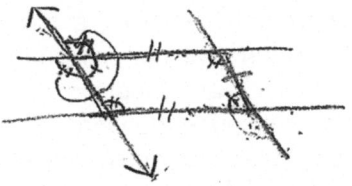

Sam drew a parallelogram first, and extended sides of the parallelogram later

Sam: This is a parallelogram because I drew it so that this side would be parallel to this side [pointing at the two longer sides of the parallelogram]. And this side would be parallel with this side [pointing at the two shorter sides of the parallelogram]

Sam provided another drawing of a different parallelogram than the first one, while she provided the following responses.

Researcher: Why is this a parallelogram?

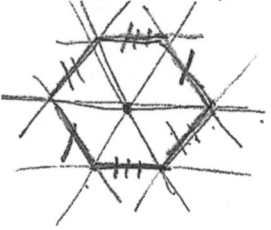

Sam drew a hexagon first, and she extended sides of the hexagon later

Sam: I think it's a parallelogram because all the sides are parallel to another side.
Researcher: Why is this a different parallelogram?
Sam: It's different because there are more sides and because the angles are different.

The preceding conversations present an interpretive description of *parallelogram* when Sam used that word at the pre-interview. With an understanding of a parallelogram being "any figure that has at least one pair of parallel sides," Sam believed "a trapezoid [a right trapezoid] is considered a parallelogram." She defined a parallelogram as, "a figure with all sides being pairs of parallel line segments," which was inconsistent with her verbal statement of identifying a trapezoid as a parallelogram. Neither Sam's written narrative nor her verbal narrative about parallelograms mentioned a necessary condition of a parallelogram being a quadrilateral. To Sam, nothing in the word "parallelogram" suggested it should be limited to 4-sided polygons. Because of that, Sam chose a hexagon as an example of a different parallelogram. When identifying and defining parallelograms, Sam focused on the necessary condition of parallel sides. In the pre-interview, Sam's understanding of parallelograms was specified as she expressed, "I actually don't know if parallelograms are strictly four-sided figures... or many shapes should be parallelograms." Table 4.19 summarizes Sam's definitions of parallelograms in the pre-interview.

Table 4.19. Sam's Definitions of Parallelograms in the Pre-Interview

Name	Definitions	Sample Polygons
Square	A four-sided figure with four right angles where all side length are equal.	 U R
Rectangle	A four-sided figure with two sets of parallel sides and four right angles.	 M F
Parallelogram	A figure with all sides being pairs of parallel line segments.	 J
Trapezoid	A four-sided figure with one set of parallel sides, with two adjacent obtuse angles and acute angles.	 N H
Rhombus	A four-sided figure with at least one set of parallel sides.	 Z L

Sam's use of the word *parallelogram* signifies a collection of polygons that share a common property of parallel sides. Based on Sam's definition, this collection of polygons could include quadrilaterals that have one pair of parallel sides such as *trapezoids*, two pairs of parallel sides such as *parallelograms*, or polygons that have more than two pairs of parallel sides such as *hexagons*. In her groupings, Sam identified the following polygons as parallelograms:

Researcher: What are the parallelograms here?

Sam: These ones are the ones [H, Z, J, L, and P] have two sets of parallel sides, but don't have the right angles, you know they have a set of obtuse angles and a set of acute angles, so they are parallelograms.

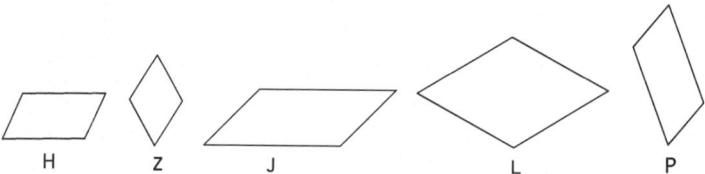

Sam did not include rectangles and squares as parallelograms, but considered them as a separate group of figures that have right angles. Sam's use of the word *parallelogram* in the pre-interview is illustrated in Figure 4.32.

Parallelograms

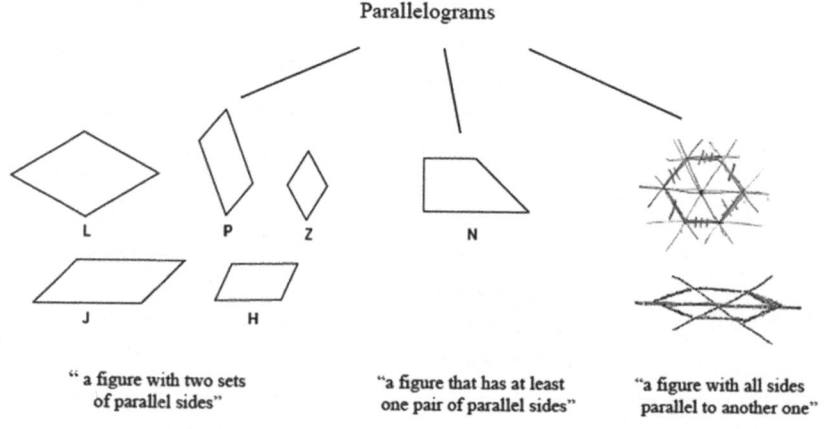

" a figure with two sets of parallel sides"   "a figure that has at least one pair of parallel sides"   "a figure with all sides parallel to another one"

Figure 4.32. Sam's use of the word parallelogram in the pre-interview.

In contrast, Sam's use of the *parallelogram* changed in the post-interview. Although Sam showed very similar routine procedures when identifying geometric figures in both interviews, her concept of a *parallelogram* was different from that of the pre-interview. For example, when Sam was asked to draw two different parallelograms, she started to draw a parallelogram and then a square.

Researcher: Why is this a parallelogram?

Sam's drawing of a parallelogram

Sam: Because it has four sides and each opposing side is parallel to one another.

Researcher: Why is this a parallelogram?

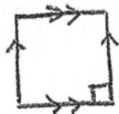

Sam's drawing of a different parallelogram

Sam: It's a square, it has four sides of equal measure and all angles are 90 degrees.

Researcher: Why is this a different parallelogram?

Sam: This one is different because all the angles and sides in this figure are equal.

Sam's use of the word *parallelogram* changed with regard to this added necessary condition of "four-sided" figure, and another necessary condition of "parallel sides." She also considered rectangles and squares as figures with 90-degree angles and as parallelograms. At the post-interview, Sam's use of the word *parallelogram* signified a collection of figures sharing common properties of being "a four-sided figure with two sets of parallel sides." She did not provide any explicit information about how these parallelograms were connected. For example, Sam grouped rhombi together with parallelograms because all rhombi have two sets of parallel sides; Sam defined a rhombus as a "four-sided figure with all side length equal in measure" but made no connections between rhombi and squares. Moreover, Sam did not mention any relations between squares and rectangles other than that they have four right angles. Therefore, it can be concluded that Sam had a good grasp of the concept of parallelograms in general, but her understanding of the hierarchy of parallelograms was not clearly demonstrated in the post-interviews. Figure 4.33 illustrates Sam's understanding of definition of a *parallelogram* in the post-interview.

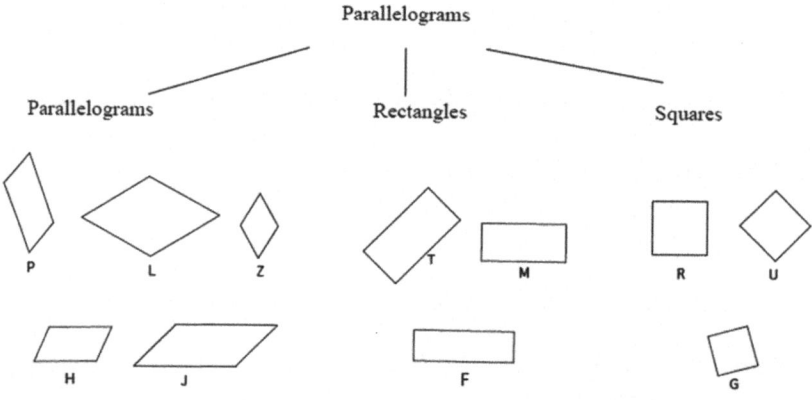

Figure 4.33. An illustration of Sam's use of the word parallelogram in the post-interview.

The changes in Sam's geometric discourse did not contradict the findings in her van Hiele geometric pre-and post-tests, but provided evidence of what might have been missed from the tests. The analysis of Sam's geometric discourses revealed a dynamic change in her geometric thinking and the way she perceive these geometric figures, when the test results suggested that she stayed at Level 2 thinking.

4.2.5 Case Five: Changes in Judi's Geometric Discourse

Judi was a second-year college student at the time of the interviews. She had taken her last geometry class in 9th grade, about five years prior to the study. The van Hiele Geometry pre- and post-test placed Judi at Level 3, a level wherein students start to reason deductively. The van

Hiele test results suggested there was no change in Judi's geometric thinking; however, the findings of Judi's geometric discourse showed the changes. A summary of changes in Judi's geometric discourse is as follows:

- Judi changed her routines of sorting polygons, from focusing on the names and attributes of the quadrilaterals at the pre-interview, to demonstrating quadrilaterals were connected with a hierarchy of the classifications of the quadrilaterals at the post-interview.
- Judi's routine of substantiation changed from using a combination of recalling and measuring routines to verify claims at the pre-interview, to routine procedures using mathematical theorems to construct new endorsed narratives the post-interview. In particular, Judi applied congruence criterions such as angle-side-angle, side-side-side for verification of congruent triangles; and used the dissection method to verify the sum of the interior angles of parallelograms in the post-interview.
- Judi used more mathematical terms and was able construct more sophisticated mathematical proofs in the post-interview.

Changes in Judi's Routines

In the pre-interview, Judi grouped eighteen polygons into groups of *triangles* (K, W, S, and X), *rectangles* (M, F and T), *squares* (U, G and R), *parallelograms/rhombi* (P, H, J, L, and Z), *quadrilaterals* (N and Q), and *other* (V).

Judi: This group is triangles:

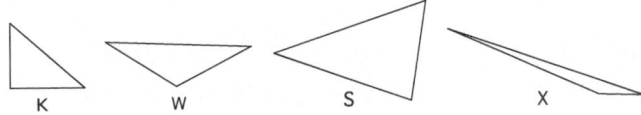

Judi: These ones are rectangles [M, F, G].

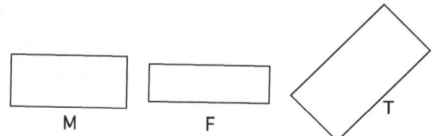

Judi: U, G and R are squares.

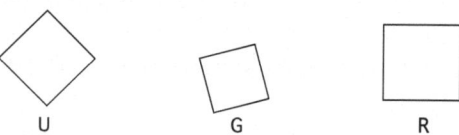

Judi: These P, L, J, Z, and H are parallelograms and rhombuses.

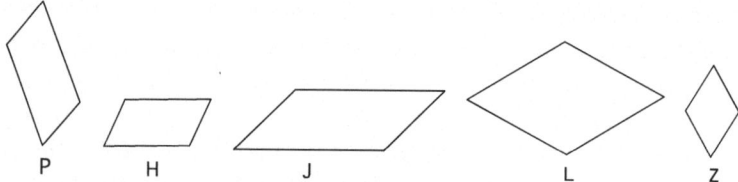

Judi: N and Q just are quadrilaterals.

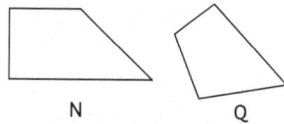

Judi: then V is just a weird shape.

Judi grouped all 3-sided polygons together and called them *triangles*; she also identified all *rectangles* together as well as *squares*. She grouped *parallelograms* and *rhombi* together, the only group with two names. Judi put a right trapezoid (N) and a quadrilateral with no parallel or equal sides (Q) together because both have just four sides. A hexagon (V) was grouped and named as "other" because "it is a just weird shape."

The researcher asked Judi to regroup the polygons differently, and she combined rectangles and squares together and called it *rectangles*, and then split the parallelgrams into paralleograms and rhombuses.

Judi: These [U and M] are just rectangles because squares can be also rectangles.

Judi: And then, L and Z are rhombuses.

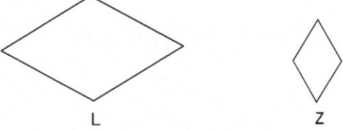

Judi: Those are *parallelograms*.

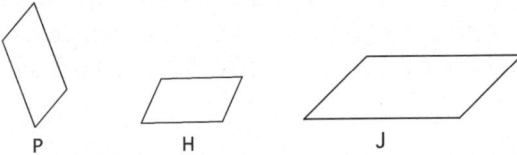

In a different approach, Judi grouped the polygons based on their angles. She grouped the polygons into a *right-angled shape* group, consisting of all polygons with at least one right angle and an *obtuse triangle* group containing all triangles with at least one obtuse angle.

Judi: These are right-angled shapes. This is a right triangle [K]; this has a right angle here [N].

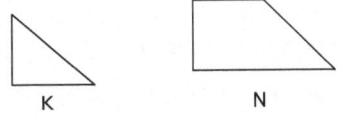

Judi: All the squares and rectangles [U, M, F, G, T, and R]

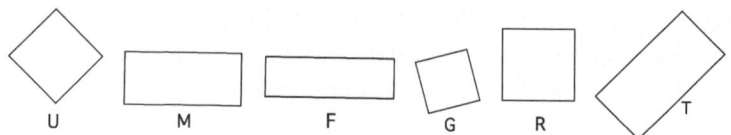

Judi: These ones I put as obtuse triangles

When the researcher asked for what happened to the remaining three polygons (S, Q and V), the following conversation took place:

Researcher: Why doesn't this triangle [S] go with any other groups?

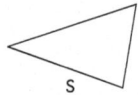

Judi: It doesn't have a right angle; it's not four-sided. So it's not a square, a parallelogram, and then ...it's not obtuse either.

Researcher: How about this one [Q]? Why is it left out?

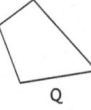

Judi: It doesn't have parallel sides and it doesn't have a right angle.

Researcher: What about this one [V]?

Judi: That's just a weird shape.

Researcher: Can this one [L] and this one [U] be grouped together?

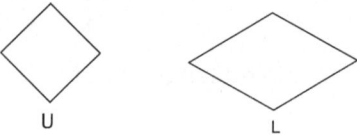

Judi: Yes. Because, they van be both seen as rhombuses.
Researcher: Why do you think they are rhombuses?
Judi: Because all their sides are equal in length.

Judi's routines of sorting polygons first focused on their common descriptive narratives (i.e., definitions), when she grouped them as triangles, rectangles, squares, parallelograms and rhombuses. She also regrouped them by the common attributes they share such as right angles, obtuse angles, or polygons having equal sides. Judi did not use any measurement tools to check the angles or the sides of any polygons for verification. Therefore, her judgments about the attributes of the angles and sides were intuition as she used the term "be seen as." Figure 4.34 summarizes Judi's routine procedures for sorting polygons in the pre-interview.

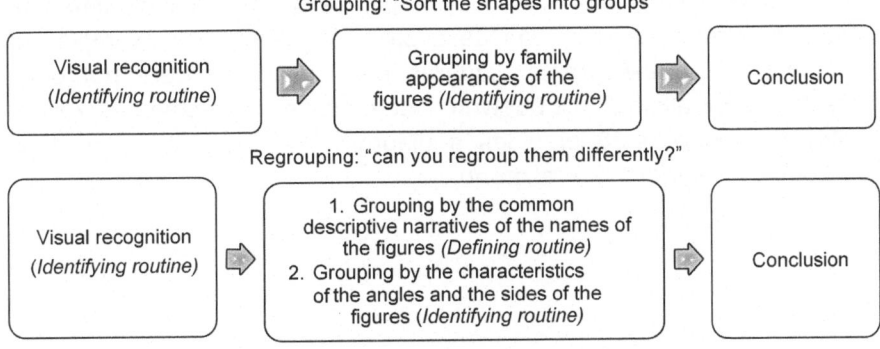

Figure 4.34. Judi's routine procedures for sorting polygons in the pre-interview.

In contrast, Judi's routines of sorting quadrilaterals showed that they were arranged with a hierarchy of classifications. Judi first grouped eighteen polygons by the numbers of their sides as *triangles*, *quadrilaterals* and a *six-sided figure*. She then divided the *quadrilateral* group into subgroups of *squares*, *rectangles*, *parallelograms*, *trapezoids*, and *quadrilaterals* (see Figure 4.35). For example, Judi identified squares (U, G and R) and then rectangles as a collection of both rectangles (M, F and T) and squares. This same pattern was observed when Judi identified parallelograms, as a group of all quadrilaterals with squares, rectangles, rhombuses and parallelograms.

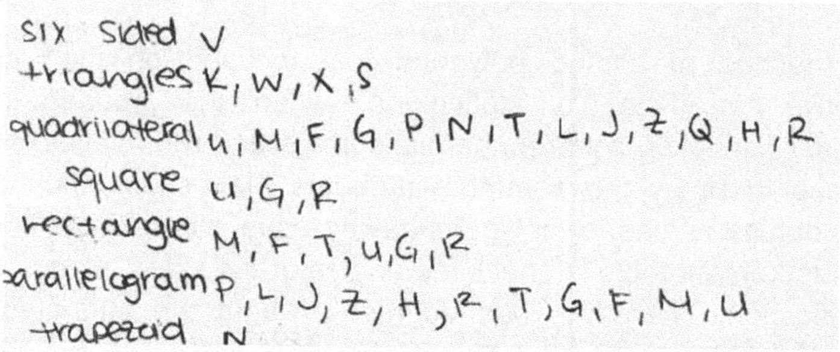

Figure 4.35. Judi's responses of grouping the eighteen polygons in the post-interview.

In the post-interview, Judi placed some parallelograms multiple times in several subgroups by common descriptive narratives. For example, in the family of parallelograms, squares (U, R and G) were not only classified as parallelograms but were also identified as rectangles. As an illustration, Figure 4.36 provides a diagram that illustrates how Judi subgrouping the *quadrilateral* group.

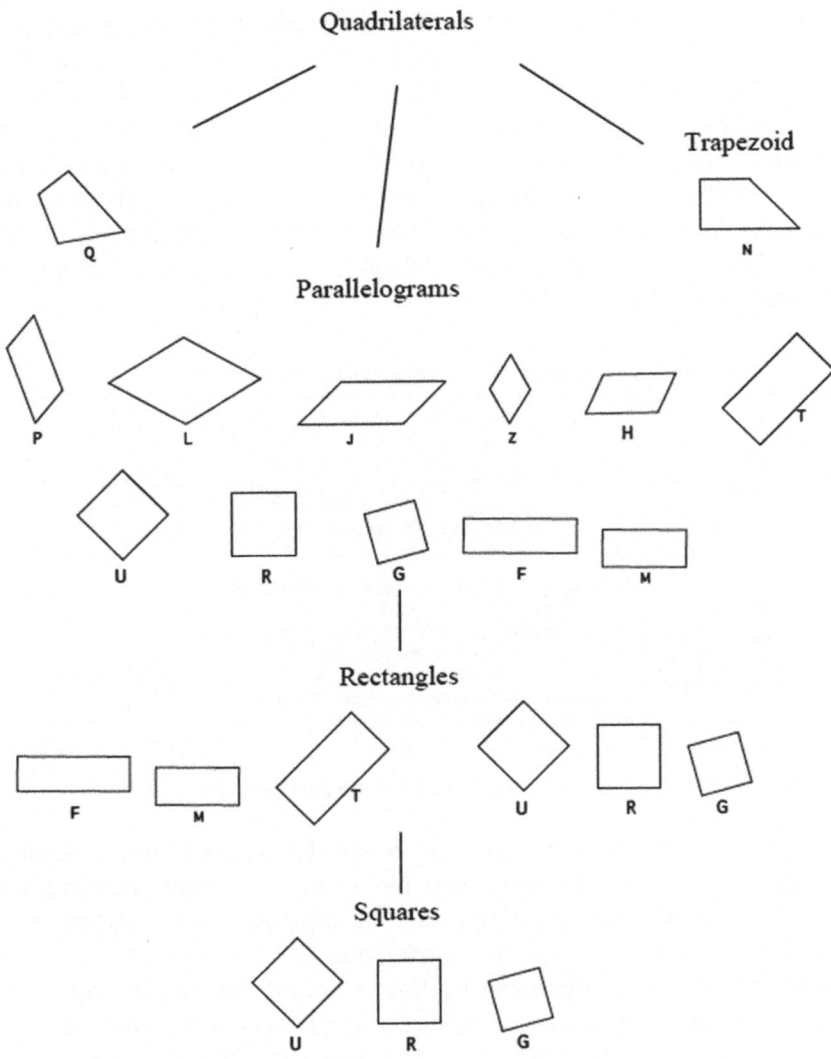

Figure 4.36. Judi's subgroups of the quadrilaterals in the post-interview.

When Judi was asked for regrouping, she regrouped triangles according to the characteristics of their angles, splitting the *triangle* group into three subgroups consisting of *obtuse triangles*, a *right triangle* and an *acute triangle*. For the *parallelograms*, Judi combined rhombuses and the squares, and called it the *rhombuses* group. When the interviewer asked Judi why she made this change, she responded, "they are all rhombuses because they have equal sides." Judi did not use measurement tools, and her routines of sorting polygons in the post-interview are summarized in Figure 4.37.

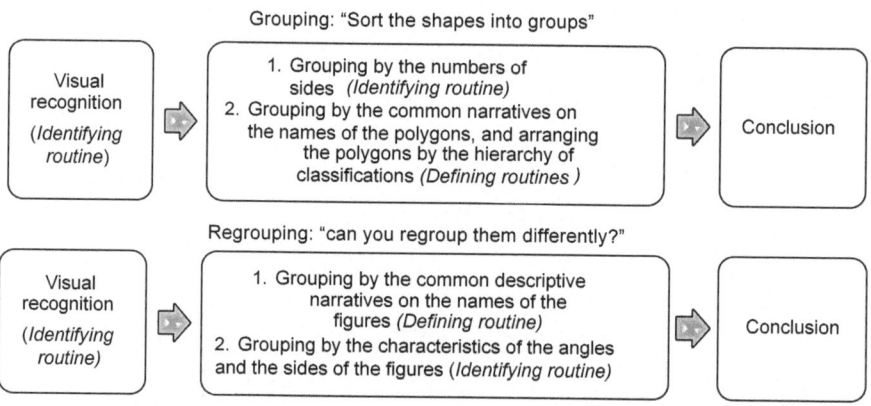

Figure 4.37. Judi's routines of sorting polygons in the post-interview.

For both interviews, Judi was able to use definitions to group eighteen polygons. She also demonstrated her ability to classify quadrilaterals into a hierarchy at the post-interview. In particular, Judi focused on the necessary conditions of the definitions in the pre-interview, and quadrilaterals were organized by their common names. In contrast, at the post-interview, Judi focused on both necessary and sufficient conditions of the definitions, and thus, the quadrilaterals were grouped with a hierarchy of classifications. Changes in Judi's geometric discourse also occurred in her routines of substantiation and these changes are reported in the following section.

Judi showed her ability to use endorsed narratives to construct new narratives at the post-interview, whereas she depended on recalling and measuring routines in the pre-interview. A recalling routine is a course of action using previously endorsed narratives, and is more about remembering what one learned previously. In this study, a measurement routine is a set of repetitive actions where participants measure the parts of polygons, and use these measurements in their identifying, verifying and substantiating processes. In the next example, Judi tried to verify that all angles in a parallelogram add up to 360 degrees.

Judi was asked to draw two different parallelograms and to discuss their angles. She drew a parallelogram and wrote that all of the angles added together equal 360° (see Figure 4.38).

A. Draw a __parallelogram__ in the space below.

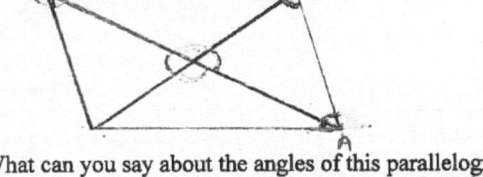

1. What can you say about the angles of this parallelogram?

the opposite angles are equal.
Angles A and B added together equales 180°
All of the angles added together equals 360°

Figure 4.38. Judi's drawing of a parallelogram in the pre-interview.

When asked for substantiation, Judi first showed that two adjacent angles added together equal 180°, and next used that result to justify her claim that "all of the angles added together equal 360°."

Judi: If you add this angle [∠A] and this angle [∠B] together, it's equal 180°.

Researcher: Why is that?

Judi: There is a name for it. I forget the term. I think this angle [∠B] is supposed to equal this outside angle here.

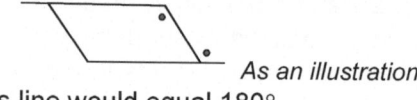
As an illustration

Judi: …and this line would equal 180°.

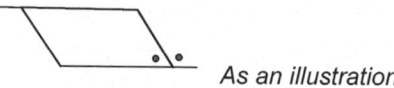
As an illustration

Researcher: How do you know these angles are equal?

Judi: I learned at school, but I don't know how to explain it. If that equals 180° added together, then this would equal 180° added together.

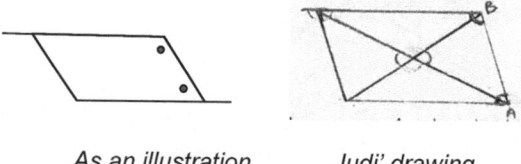

As an illustration   *Judi' drawing*

Judi: So all the angles [in the parallelogram] together would equal 360°.

In this excerpt, Judi used recalling routine to conclude that two angles form a straight equal to 180°, without knowing why the rule worked in a given situation. For example, in looking at the two adjacent vertex angles in a parallelogram, Judi concluded that they added up to 180°. This conclusion is correct because we know by definition that a parallelogram has opposite sides parallel, and we can use propositions about parallel lines and their transversals to conclude that the two angles add up to 180°. Similarly, Judi was able to recognize that the alternating interior angles were congruent as a fact without knowing it as a consequence of two parallel lines cut by a transversal. Note that Judi utilized logical thinking in the statement, "*if* that equals 180…*then* all the angles…" Judi provided similar statements about the angles of a parallelogram in the post-interview (see Figure 4.39).

A. Draw a _parallelogram_ in the space below.

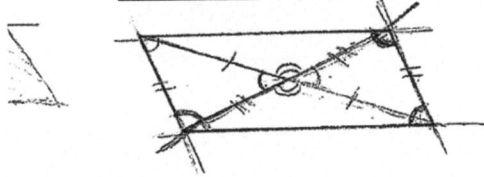

1. What can you say about the angles of this parallelogram?

the opposite angles are equal
adjacent angles equal 180°
all angles added together equal 360°

Figure 4.39. Judi's drawing of a parallelogram in the post-interview.

In contrast to her response at the pre-interview, Judi substantiated the statement of two adjacent angles added up 180° at the post-interview.

Judi: These are adjacent angles.

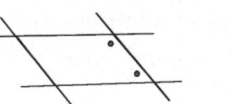 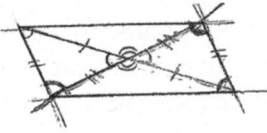

As an illustration   *Judi's drawing*

Judi: These are alternate interior angles and they are equal, because by the parallel lines. Then this plus this angle is 180° because angles on a line.

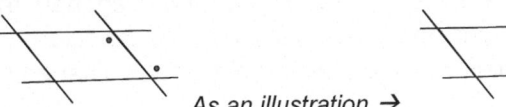

As an illustration →

Judi: And so this and this, the adjacent angles equal 180°.

 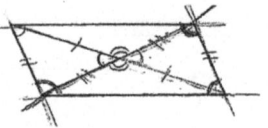

As an illustration   *Judi's drawing*

In this excerpt, Judi began to use mathematical terms to express her ideas. She used the words "adjacent angles" and "alternating interior angles" to replace her informal use of "this angle" and "that angle," and "angle inside" and "angle outside" at the pre-interview. More importantly, Judi substantiated her claims using endorsed narratives. For example, when she produced the utterance that alternate interior angles were equal, she used a phrase "*by … the parallel lines*"; and when Judi declared another utterance that two angles add up to 180°, she used the phrase "*because angles on a line*." When the researcher asked Judi why she thought all angles in a parallelogram added up to 360°, the following conversation took place:

Researcher: What are the angles that add together to equal 360°?
Judi: All the interior angles [vertex angles], this plus this plus this plus this equal 360°.

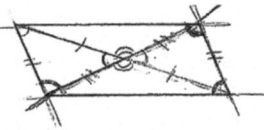

Researcher: Why is that?
Judi: Because if you draw a line, the diagonal, there are two triangles, and the interior angles of a triangle equal 180°. So, two triangles would equal 180° plus 180°. That equals 360°.

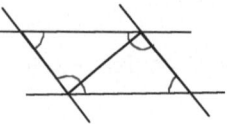 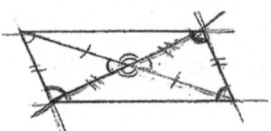

As an illustration   *Judi's drawing*

Judi used dissection method to draw a diagonal that cut the parallelogram into two triangles, and applied the endorsed narrative that the sum of the interior angles of a triangle equals 180° to complete her justification that all interior angles in a parallelogram added up to 360°.

Judi's routine procedures also changed in verifying the congruent parts of a parallelogram. She was aware of the abstraction of congruent parts of the parallelograms at the post-interview, whereas she only used measurement routines to check the congruent parts at the pre-interview. For example, in the pre-interview, Judi was prompted to substantiate her claim about the diagonals of a square, and she showed that the diagonals of the square were equal by checking the measurements of the diagonals.

> Researcher: What can you say about the diagonals of this parallelogram [a square]?
> Judi: They're equal in lengths.
> Researcher: How do you know?
> Judi: Because they are all equal sides [Pointing at the sides of the square]. I am sure they would be all equal diagonals. I can check.
> Judi: Let's do it in inches. This one [one diagonal] is 2.7. That's also 2.7 [another diagonal].
> Judi: Yeah, diagonals are equal.

Judi first made a claim about why the diagonals would be equal, and then used a ruler to measure the length of the diagonals, getting measurements for each diagonal of 2.7 inches. With this confirmation, Judi concluded that the diagonals were of equal length. Judi then made another statement about the diagonals of a square being perpendicular to each other.

> Researcher: What can you say about the diagonals of this parallelogram [a square]?
> Judi: I think they're perpendicular to each other.
> Researcher: What do you mean when you say perpendicular?

Judi: At the intersection, they create a 90° angle.

Researcher: How do you know they are 90° angles?

Judi: I can measure it… Yeah… it's 90°.

Researcher: How do you know this is true for all squares?

Judi: I am pretty sure that they are all 90°. This is 90°, and that's 90°. They are all equal.

When Judi was asked, "how do you know that they are all 90°?" her course of action focused on the concreteness of the congruent angles, in using a protractor to check the angles and using measurement routines to verify her claims.

Judi declared the same narrative about the diagonals of a square, stating that the diagonals were equal length in the post-interview. However, when asked for substantiation, Judi responded with "I can prove again that the triangles are congruent" because she had just substantiated that the diagonals in a rectangle were equal, so she applied that argument. As an illustration, Table 4.20 summarizes Judi's routines of substantiating the two diagonals of a rectangle were equal with corresponding transcript.

Table 4.20. Judi's Routine of Substantiation for "Two Diagonals Are Equal" in the Post-Interview

Routine Procedures	Transcripts
1. Identify two congruent triangles	Judi: this triangle and this triangle are congruent [pointing at the shaded triangles]:
2. Declared narrative	Judi: and then that side equals this side [pointing at the diagonals]:
Prompt for verification	*Researcher: How do you know these two triangles are congruent?*

3. Verification of two congruent triangles	
3.1 Identify corresponding angles of the triangles	Judi: well, these are 90°[pointing at the two right angles] 
3.2 Identify corresponding sides of the triangles	Judi: they have a common side so that would be the same for both triangles [Marked a tally on the common side]. 
3.3 Identify another corresponding sides of the triangles	Judi: opposite sides that are parallel are equal [Marked two tallies on the opposite sides of the rectangle] 
4. Verify congruent triangles using S-S-S correspondence	Judi: that gives you side, angle, side and makes these two triangles congurent.
5. Conclusion	Judi: by that, you can conclude that this side and this side are equal [pointing at the diagonals of the rectagnle].

Table 4.20 shows that Judi first identified a pair of congruent triangles with diagonals as one set of corresponding sides of the triangles, and drew a conclusion about the equivalence of the diagonals. After the interviewer prompted for substantiation, Judi provided a sequence of steps of selecting three elements needed for the verification of congruent triangles. During this selection, Judi did not use a ruler or protractor to check the measurements of the sides and angles, but instead used identifying routines and defining routines to draw conclusions. In addition, she used the definition of a rectangle to identify two corresponding angles that were 90° using a definition of rectangle.

Researcher: Why is this a parallelogram?

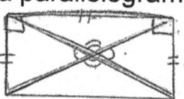

Judi: Because opposite sides are parallel and the opposite sides are equal.

Researcher: Why is this a different parallelogram from the one you drew?

Judi: Because they all form 90° angles, all the angles are equal, not just the opposite angles.

Judi: This is a rectangle.

Judi used an identifying routine to identify the right angles and opposite sides as parts of the rectangle, then used a *defining routine* to confirm her choice of the elements of the congruent triangles needed for verification. Because the polygon was a rectangle, all angles were equal and opposite sides were equal. Similarly, Judi applied this proof in the case of a square in the next example, as she made the connections between a square and a rectangle.

Judi stated, "diagonals [of a square] are perpendicular to each other," "diagonals bisect each other," and "diagonals bisect the angles". As noted earlier, Judi made the connection between the diagonals of a square and the diagonals of a rectangle, as in both cases their "diagonals were equal." Later she made another connection between the diagonals of a square and the diagonals of a parallelogram, as in both cases their "diagonals bisect each other." Judi used the Angle-Side-Angle criterion to substantiate two congruent triangles in a parallelogram, and applied that result to draw the same conclusion in the case of a square. The following excerpt provides some insight of Judi's substantiation of "diagonals are perpendicular to each other":

Judi: They are perpendicular to each other. [Pointing at the intersection of the diagonals]:

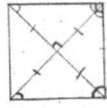 *Judi's drawing*

Researcher: How do you know?
Judi: Because the angles of the square are 90°

 As an illustration

Judi: And the diagonal cut this angle in equal halves. [Pointing at the
two angles]

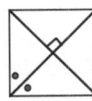 *As an illustration*

Judi: So this would be 45° and this would be 45°. [Writing 45° at
each angles]

 As an illustration

Judi: So that would equal to 90° [Adding the two 45° angles]
Judi: For angles a triangle it would be 180 total, so it would have to
be 90°. [Pointing at the intersection of the diagonals]

Judi used the fact that the diagonals form 90° angles to verify that the
diagonals were perpendicular to each other. In the process of this
verification, she made one assumption, that "the diagonal cuts this angle
in half," which was a new statement. Therefore, the researcher asked for
substantiation:

Researcher: How do you know that the diagonal cuts the angle in
two halves?

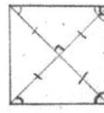 *Judi's drawing*

Judi: Because, if the diagonals are bisecting each other, then the halves are all equal lengths because the diagonals are equal lengths [pointing the halves of one diagonal]

Judi: So from there [Pointing at the base angles of the triangle], if these sides are equal, and it would be an isosceles triangle. By the definition of an isosceles triangle, these angles would have to be equal.

 As an illustration

Judi: And then the same with this triangle [Pointing at the adjacent triangle], it would also be an isosceles.

Judi: So, these are the same isosceles triangles. So, the triangles are all congruent, and the angles would all have to be congruent.

 As an illustration

Judi: But this is 90°. So, if these angles are equal... then they are all 45°. So the diagonals cut the angles in half.

 As an illustration

Judi applied several endorsed narratives to substantiate her declared narrative "the diagonals cut the angles in half." She first used "diagonals bisect each other" and "diagonals are equal," both newly endorsed, and to conclude that "the halves are all equal." She used this newly endorsed narrative to identify congruent isosceles triangles. By using the properties of isosceles triangles, Judi showed that all corresponding [base] angles were equal. Knowing that the figure was a square, with 90° angles, she concluded that the diagonals cut the angles in half.

In summary, Judi changed her routines of substantiation, as her geometric thinking moved away from an object level of measuring and checking the congruent parts of the parallelogram at the pre-interview, towards an abstract level of substantiation, using endorsed narratives (i.e., congruence criterions, theorems, etc.) to verify her claims at the post-interview.

Changes in Judi's Word Use
Judi was able to use more mathematical terms (e.g., alternating interior angles, bisect, perpendicular, etc.) to describe her geometric thinking in the post-interview than she did in the pre-interview. She also demonstrated deeper understandings of how quadrilaterals and parallelograms were connected during the post-interview. In this section, findings of the changes in Judi's word use are provided. Judi's use of the words such as *quadrilateral, parallelogram, rectangle, square, rhombus, trapezoid,* and *kite* in both interviews are summarized in Table 4.21.

Table 4.21. Frequencies of Judi's Use of the Names of Quadrilaterals in the Two Interviews

Name	Frequency				Total Frequency	
	Pre-T1	Post-T1	Pre-T2	Post-T2	Pre	Post
Quadrilateral	3	4	0	0	3	4
Parallelogram	5	3	1	4	6	7
Rectangle	5	2	0	1	5	3
Square	6	1	1	3	7	4
Rhombus	5	3	0	1	5	4
Trapezoid	3	2	1	0	4	2
Kite	0	0	0	0	0	0

During the interviews, the *parallelogram* (n=13) was motioned most, and the word *square* (n=11) was the second most frequently used, and *rectangle* (n=8) was third. The word *kite* was not mentioned at all in both interviews, and *trapezoid* (n=6) was the second least mentioned. There was an increase in use of the words *parallelogram*, *square* and *rhombus* in the post-interview, and use of the word *rectangle* doubled in the post-interview. Judi's use of the names of quadrilaterals was much lower than other pre-service teachers' use of those names.

In an earlier section, Judi's routines for sorting quadrilaterals were described. For example, at the pre-interview, Judi first grouped the figures by their names and then by the characteristics of their angles. However, she was confused about how a *trapezoid* and a *parallelogram* were related. Such confusion was detected when Judi drew a parallelogram and then drew a trapezoid as a different parallelogram (see Figure 4.40).

A. Draw a <u>parallelogram</u> in the space below.   space below, draw a parallelogram

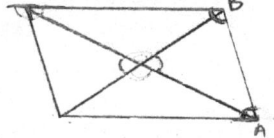 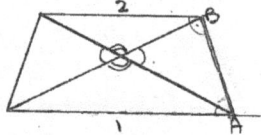

Figure 4.40. Judi's drawing of two different parallelograms in the pre-interview.

When the researcher asked Judi to draw a new parallelogram that was different from the one she drew previously, she provided the following responses:

Judi: I don't know if this is right, but I am going to draw it.

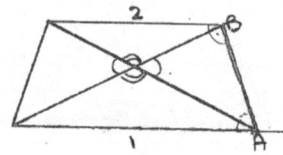

Judi: I think this is wrong.

Researcher: Why? I just want to know what bothers you.

Judi: I know these are parallel sides [pointing at the two parallel sides], but I don't know if a trapezoid is also a parallelogram. I am not sure.

Researcher: What is a trapezoid?

Judi: I am not sure, cause I think trapezoids need two parallel sides.

Researcher: What is a parallelogram?

Judi: A four-sided shape with like this [Pointing at her first drawing], opposite sides that are parallel?

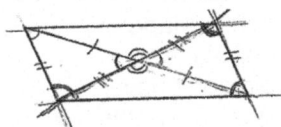

The preceding excerpt shows that Judi was unclear about what a *trapezoid* was, as she explained, "I don't know if a trapezoid is also a parallelogram". She produced an utterance stating that a trapezoid needs two parallel sides. In her definition, Judi made no distinction between a trapezoid and a parallelogram in her definition in the pre-interview. Table 4.22 summarizes Judi's definitions of parallelograms in the pre-interview.

Table 4.22. Judi's Definitions of Parallelograms in the Pre-Interview

Name	Definitions	Sample Polygons
Square	A four-sided shape with equal length on all sides and perpendicular diagonals of equal lengths.	U R

Rectangle	A four-sided shape with opposite sides of equal length with equal diagonal lengths	![rectangles labeled M and F]
Parallelogram	A four-sided shape with opposite sides parallel to each other.	![parallelograms labeled J]
Rhombus	A four-sided shape with equal sides, diagonals don't have to be equal in lengths.	![rhombi labeled Z and L]

In the pre-interview, Judi first included parallelograms and rhombi in the *parallelogram* group and later split it into parallelograms and rhombi as two subgroups. Similarly, she included rectangles and squares in the *rectangle* group, and then divided them into a *rectangle* group and a *square* group. Judi did not make connections between the parallelograms and rectangles, but she did consider a square was a rhombus after she was prompted.

Researcher: Can I group this one (L) and this one (U) together?

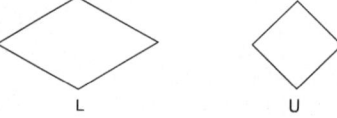

Judi: Yes.
Researcher: Why is that?
Judi: Because they both are seen as rhombuses.
Researcher: Why do you think they are rhombuses?
Judi: Because all their sides are equal length.

In a different task, Judi was presented with pictures of a rectangle and a square, and was asked to identify if they were parallelograms. Judi responded, "yes" and explained, "they are parallelograms, because they have both opposite sides parallel." Judi used the word *parallelogram* to describe a collection of quadrilaterals having parallel sides. In particular, she referred the word parallelogram to two branches of quadrilaterals: *parallelograms* and *rectangles*, while her understanding of the hierarchy of parallelograms was not well demonstrated in the pre-interview. Figure 4.41 illustrates Judi's use of the word parallelogram in the pre-interview.

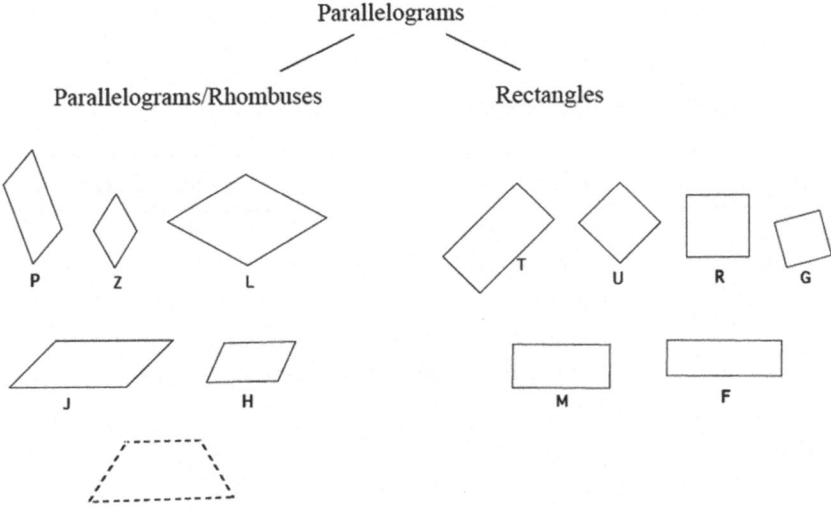

Figure 4.41. Judi's use of the word parallelogram in the pre-interview.

In contrast, when Judi used names of parallelograms in the post-interview, it showed a hierarchy of classifications among the parallelograms. For example, she used the word *parallelogram* as a description of quadrilaterals with two pairs of parallel sides. Among the parallelograms, Judi also recognized two subgroups: a *rectangle* group consisting of parallelograms with right angles, and a *parallelogram* group consisting of rhombuses and parallelograms.

Judi's drawings of two different parallelograms are shown in Figure 4.42, and her definitions of quadrilaterals are summarized in Table 4.23.

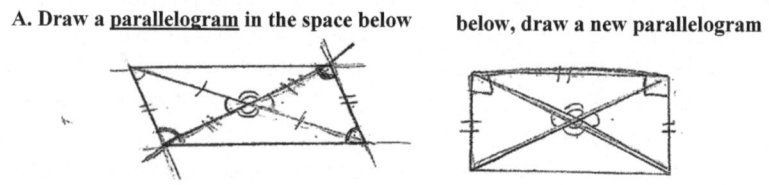

A. Draw a <u>parallelogram</u> in the space below   below, draw a new parallelogram

Figure 4.42. Judi's drawings of two different parallelograms in the post-interview.

Table 4.23. Judi's Definitions of Parallelograms in the Post-Interview

Name	Definitions	Sample Polygons
Square	A four-sided figure with all equal sides and all 90° angles.	U
Rectangle	A four-sided figure with opposite sides equal and all 90° angles.	M
Parallelogram	A four-sided figure with opposite sides equal and parallel.	J
Trapezoid	A four-sided figure with one pair of opposite parallel sides	N
Rhombus	A four-sided figure with all equal sides and opposite sides parallel.	Z

Based on the characteristics of the sides, Judi split the *rectangle* group into *squares*, a group of rectangles with all equal sides, and *rectangles*. Judi then grouped *rhombi* and *squares* together because all their sides were equal. Figure 4.43 illustrates how parallelograms are connected within a hierarchy during the post-interview. In this hierarchy, the word *parallelograms* denoted a collection of quadrilaterals with different *names*, and these names were *parallelograms*, *rectangles*, *squares,* and *rhombi*. Although these *names* identify polygons with different physical appearances and attributes of their angles and sides, they are all *parallelograms.*

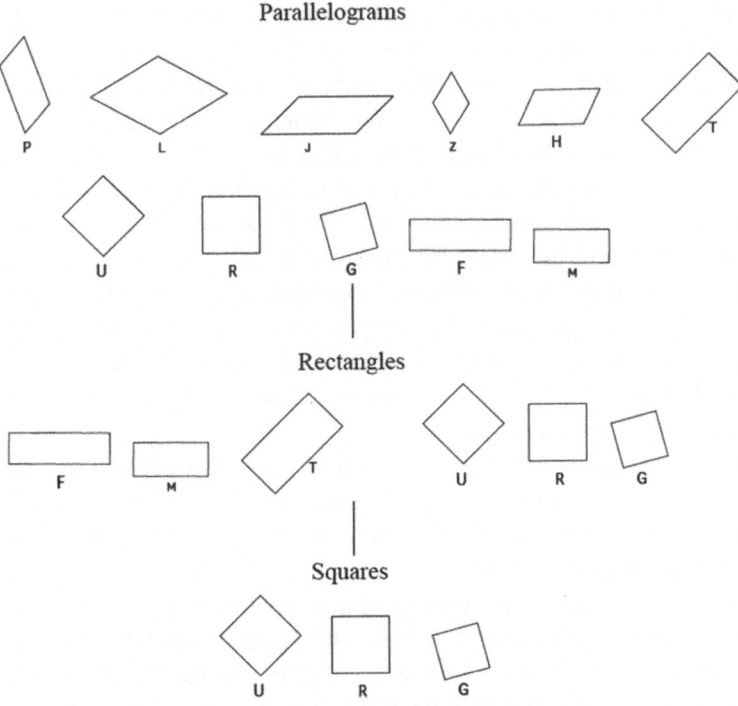

Figure 4.43. Judi's use of the word parallelogram in the post-interview.

Although the van Hiele geometry pre- and post-test results (Level 3-Level 3) showed no change in Judi's geometric thinking, there were changes in her geometric discourse. These changes indicated she was making transitions from informal mathematics discourse towards more sophisticated mathematical discourse. For example, she used more mathematical terms at the post-interview, and used deductive reasoning to produced narratives with justifications such as "those sides are equal *by corresponding parts in congruent triangles,*" "angles are equal *because vertical angles are equal*" and "these sides are equal *by the definition of a parallelogram.*"

4.3　The van Hiele Levels as Geometric Discourse

The van Hiele geometry pre- and post-tests suggest changes in Molly (Level 1-Level 3), Ivy (Level 0-Level 3) and Kevin's (Level 3-Level 4) geometric thinking, and indicated no change in Sam (Level 2-Level 2) and Judi's (Level 3-Level 3) geometric thinking. The van Hiele geometry test results were general and descriptive but static in nature, as they provided less detail about geometric thinking and lack of details on the development of geometric thinking. Using a discourse approach to analyze thinking shows the dynamic changes in pre-service teachers' geometric thinking. In particular, this discourse approach provides evidence on how geometric thinking is developed and how thinking is different from person to person, with a particular focus on these pre-service teachers' use of mathematical words and routines when engaged in particular tasks.

In the previous section, I described in detail the changes in five pre-service teachers' geometric discourse. In the following section, I summarize the results of changes in each pre-service teacher's geometric discourse.

Ivy moved three van Hiele levels from Level 0 to Level 3. Ivy's geometric discourse changed from grouping quadrilaterals by their visual appearance, to classifying quadrilaterals with a hierarchy of classification using definitions. In addition, Ivy's substantiation routines reveal that she used endorsed narratives to construct mathematical proofs. Molly moved two van Hiele levels from Level 1 to Level 3. Molly's geometric discourse changed from identifying quadrilaterals by their visual appearance, to classifying quadrilaterals by their partial visual properties and definitions. Molly's substantiation routines were not observed at the interviews, nor did she demonstrate her understanding of constructing mathematical proofs. Kevin moved one van Hiele level from Level 3 to Level 4. Kevin's geometric discourse changed from grouping quadrilaterals by the number of sides and by their names and angles, to classifying quadrilaterals by their common descriptive narratives with a hierarchy of classifications. Kevin's substantiation routines also changed from verifying the congruent parts of parallelograms using recalling, measuring, and constructing routines, to formulating mathematical proofs using mathematical axioms and propositions.

Although the van Hiele geometry test results indicate no changes in Sam (Level 2-Level 2) and Judi's (Level 3-Level 3) geometric thinking, the discourse analysis results show the changes in their geometric discourse. Sam's geometric discourse changed from describing the processes of activities using transformations such as reflections, translations, and rotations, to using endorsed narratives to construct mathematical proofs. In addition, Sam's use of the word *parallelogram* changed from all polygons having pairs of parallel sides without considering all the necessary conditions, to identifying parallelograms as quadrilaterals who have two pairs of parallel sides without a hierarchy of classification. Judi's geometric discourse changed from focusing on the names and attributes of the quadrilaterals, to demonstrating that quadrilaterals were connected with a hierarchy of the classifications of the quadrilaterals. Judi's substantiation routines also changed from using a combination of recalling and measuring routines to verify claims, to using endorsed narratives (i.e., triangle congruent criterion) to construct mathematical proofs.

The interview data show the subtle differences in discourse between van Hiele levels (i.e., Kevin's Level 3 to Level 4), as well as differences within a van Hiele level from person to person, such as Molly, Ivy, Judi and Kevin's Level 3 geometric discourse, and from the same person such as Sam's Level 2 to Level 2 and Judi's Level 3 to Level 3 geometric discourse. I will give a thorough discussion about the results of these five pre-service teachers' van Hiele levels and their geometric discourse in Chapter 5.

The interview data of the study also provides a base for the following theoretical approach: each van Hiele level of thinking has its own unique geometric discourse. Therefore, the development of geometric thinking is a development of one's geometric discourse. This approach includes 1) translating the van Hiele levels of thinking into discursive terms using the four characteristics of geometric discourse (see Tables 2.2 and 2.3); and 2) developing a model, *the Development of Geometric Discourse,* and using interview data (see Table 4.24, on p.181).

The model, *the Development of Geometric Discourse* is compatible with the first four van Hiele levels of thinking and it consists of two components: (1) *Geometric Objects* shed lights on one's use of geometric words (as signifier), saming criterions, realizations, and systems of objects or figures; (2) *Routines* captures one's discourse specific course of actions in response to mathematical tasks which include, but are not limited to, identifying routines and defining routines. Note that *endorsed narratives* and *visual mediators* are not listed as separate characteristics in the model, because they are interwoven with *geometric objects* and *routines.* For example, narratives produced by a student through written or verbal utterance, provide contexts in which mathematical words are used and the nature of geometric figures played in that particular discourse. *Visual mediators* are collections of symbolic artifacts, geometric figures and their parts, and when a student interacts with visual mediators, such interaction provides clues on how a student's routines operate: at an object-level or an abstract-level.

In a geometric discourse, the *Geometric Object* is defined as a mathematical object, which constitutes "the thing" that we discuss in a discourse specific context. In this study, "the thing" often referred to geometric figures such as a triangle, a quadrilateral, or their parts (e.g., sides, angles, etc.). At different van Hiele levels, the same geometric figure discussed may not be the same the *geometric object* in the corresponding geometric discourse. For example, at Level 1 the word *square* is used as a label to a picture of a square, and is just a matching of a word with a shape (signifier). All squares can be grouped together because they all fit this family appearance of four sides equal with their angles and look like the "corners of a desk." However, at this level, students do not group a rhombus and a square together because they do not have the same family appearance even though they both have four sides equal (saming). All figures are grouped by their names only, because each name represents a family appearance (realization), and there is no hierarchy connecting geometric figures at Level 1 (system of objects). At Level 3, the word *square* can also represent a parallelogram, a rectangle or a rhombus because a square fits the common descriptive narratives "opposite sides parallel," "a parallelogram with four right angles," and "parallelogram with four equal sides" (saming). Thus, a square can be grouped with parallelograms, rectangles and rhombi (realization).

If a student (Molly) moves from Level 1 to Level 3 thinking, the polygons called "square" and "rectangle" play different roles as *geometric objects* in Level 1 geometric discourse and Level 3 geometric discourse. For example, at Level 1 geometric discourse, naming geometric figures as a rectangle or a square is about assigning a proper name to a figure, one is named "square", and the other one is named "rectangle." To Molly, a square and a rectangle are given two different names because of their visual appearances (one looks like a square, and one looks like a rectangle). Therefore, at Level 1 geometric discourse, *geometric objects* (e.g., squares, rectangles, etc.) are discursive objects that are used passively and are organized concretely, unstructured by their visual appearance. In contrast, in Level 3 geometric discourse, Molly not only named squares and rectangles as "square" and "rectangle" (their proper name), but also identified them as "rectangle" and "parallelogram" (their common name) by common descriptive narratives. That is, a geometric figure such as a square has multiple names, a parallelogram, a rectangle, by definitions of parallelogram and rectangle. Therefore, at Level 3 geometric discourse, *geometric objects* (e.g., squares, rectangles, etc.) are discursive objects that are object-driven, more abstract in nature, and are organized by common descriptive narratives (i.e., mathematical definitions). To conclude, a square as a geometric figure makes a general character in van Hiele levels; in this model, a square as a geometric object plays different roles at different levels of discourse.

Routines are repetitive discursive patterns that can be predicted in similar situations. In this model, routines consist of *identifying routines* and *defining routines*, which determine the role of definitions played in

Level 1 to Level 3 geometry discourse that are aligned with van Hiele levels 1 to 3. Identifying and defining routines detect ways in which a student perceives a geometric figure and her/his reasoning about it as a repetitive pattern. For example, at Level 1 geometric discourse, Molly identified a polygon as a square because it looks like (visual recognition) what she used to call it by its appearance (recalling). When asked for justification, Molly's course of action was self-evident. At Level 3 geometric discourse, Molly identified a square because it has four equal sides and right angles using a definition for squares. Molly also used identifying and defining routines to declare that rectangles and squares were parallelograms and all squares were rectangles. Therefore, the defining and identifying routines provide clues of how the student uses definitions as a course of action. This model is developed to identify changes in students' geometric discourse as they move toward a higher level of thinking. In particular, each subsequent level of discourse is a product of reflection on the geometric discourse that has been developed so far.

Table 4.24 The Development of Geometric Discourse

Level of Discourse	Geometric Objects				Routines				
	Signifier	Saming	Realization	System of objects	Identifying Routines	Examples of Justifying Identifications (Why is it x?)	Defining Routines		
							How (What is x?)	When	
1	• Proper name • Passive use	• Family appearances	• Primary d-objects	• Unstructured Concrete d-objects	• Visual recognition • Self-evident	• Because it looks like it and it is	--	--	
2	• Common name • Routine driven	• Visual properties	• Primary d-objects	• Unstructured Concrete d-objects with disjoint categories	• Visual recognition • Partial properties check	• Because I can see it • Because I measure it	• Visual properties • Recalling	Necessary conditions for words use	
3	• Common name • Object driven	• CDN on the names of the figures • CDN on the properties of figures	• Concrete d-objects • Objectification started	• May or may not have hierarchy of classifications	• Visual recognition • Definitions check • Informal deductive reasoning	• If it is a square than it has to be a rectangle • Because they have two pairs of parallel sides	• Figures & terms are describe using definitions	Necessary & sufficient conditions for words use	
4	• Common name • Object driven • Common relations among definitions and theorems, etc.	• CDN on the names of the figures • CDN on the properties of figures	• Abstract d-objects • Objectification formed	• Hierarchy of classifications	• Construction of formal proofs • Deductive reasoning with logical sequences	• The two angles formed a straight angle, so they added up to 180 degrees.	• Figures & terms are described using definitions	Necessary & sufficient conditions for words use	

Note: d-object indicates discursive-object; CDN is an abbreviation of common discursive narrative(s)

5 Chapter Five: Discussions

This study used the van Hiele model of geometric thinking and Sfard's analytical framework on mathematical discourse to describe pre-service teachers' geometric thinking and to determine the changes in their learning. The van Hiele's and Sfard's perspective of learning are described as follows: the van Hiele's (1959) view learning as moving to a higher level of thinking, and argue that a learning process is similar to the process of learning a new language. Building on the legacy of Vygotsky, Sfard (2008) views learning of mathematics as extend or modify ones mathematical discourse, and argues that a learning process is a process of becoming a participant in mathematical discourse— learning to think in a mathematical way. We are at the moment of changing our teaching practices associated with some new conceptualizations of learning. Thus, this study is intended to provide evidence to conceptualize students' mathematical thinking through their mathematical discourse, and to identify changes in their learning.

5.1 Summary of the Results

The study combined pre-service teachers' pre- and post-test results with their interview responses to investigate their geometric discourse at each van Hiele level in the context of triangles and quadrilaterals. These results reveal the multi-semiotic nature of geometric thinking and its complexity. Three conclusions have emerged: 1) When a pre-service-teacher's geometric thinking moves towards a higher van Hiele level, her/his geometric discourse also changes; 2) When a pre-service teacher's geometric thinking stays at a van Hiele level, there are changes in her/his geometric discourse within that van Hiele level; and 3) the pre-service teachers whose van Hiele level of thinking is at Level 3, have differing geometric discourses. All these changes and differences in geometric discourse are detected by pre-service teachers' use of mathematical words and their routines as a course of actions during the interviews.

5.1.1 Changes in van Hiele Levels

Changes in pre-service teachers' (n=63) competencies suggest their overall performance on the van Hiele geometry pre-test and post-test. There are improvements in answering questions related to van Hiele Levels 1 to 3 at the post-test. Illustrated by the following:

- More than ninety-five percent of the participants correctly named triangles, squares, rectangles, and parallelograms at the post-test.
- More than ninety-five percent of the participants at the post-test correctly identified the properties of isosceles triangles, squares, rectangles, and rhombi related to their sides, angles, and diagonals.
- About ninety percent of the post-test participants correctly used logical statements regarding triangles, squares, rectangles, and parallelograms.

These changes show that these pre-service teachers gained familiarity with figures like triangles, squares, rectangles, rhombi and parallelograms, and with their properties. These pre-service teachers demonstrated a greater understanding about the properties of angles and sides in parallelograms, but less on the properties of diagonals. The van Hiele geometry pre-test and post-test results suggest pre-service teachers' weaknesses in using deductive reasoning to construct proofs (Level 4) and in abstract thinking (Level 5).

It is indicated in the previous chapter that the study set out to explore participants' van Hiele levels of thinking and their geometric discourse while taking a geometry course, rather than answering the questions of "What and how did these pre-service teachers learn from the course?" and "What are the relations between their learning and changes in their thinking?" In short, the geometry course these pre-service teachers (n=63) took during the time of the study does not serve as a "treatment" of the study. However, from the van Hiele geometry pre-test and post-test results, one could assume that the experiences these pre-service teachers had in from the geometry course contributed to their learning and helped them to develop their geometric thinking from a lower van Hiele level to van Hiele Level 3. The tests results also show that a student entering the class at Level 3 likely would stay at Level 3. This is to be expected, as the course is designed for future elementary and middle school teachers, and its materials emphasized activities mostly at Levels 1 to Level 3 of geometric thinking and included only a brief introduction to constructing proofs.

The van Hiele geometry pre-test and post-test results help determine sixty-three pre-service teachers' competence in geometry and describe their geometric thinking as a whole, but the test results do not provide details on changes in their thinking at an individual level. To add more information about these pre-service teachers' geometric thinking, changes in their geometric discourse through one-on-one interactions are analyzed and compared.

5.1.2 Changes in Geometric Discourse

Using Sfard's (2008) communicational approach to investigate geometric thinking, discourse serves as the object of the mathematical activity and the unit of analysis. From this perspective, the pre- and post-interview results capture subtle differences in pre-service teachers' geometric discourse:

1) When a pre-service-teacher's geometric thinking moves towards a higher van Hiele level, her/his geometric discourse also changes;
2) When a pre-service teacher's geometric thinking stays at a van Hiele level, there are changes in her/his geometric discourse within that van Hiele level; and
3) Among the pre-service teachers whose van Hiele level of thinking is at Level 3, their geometric discourses are different.

The results confirm that these pre-service teachers have developed some concepts about quadrilaterals because their word use has changed. *Word use* focuses on geometric names of quadrilaterals (i.e., trapezoid, parallelograms, rectangles, squares and rhombuses) and their use. Therefore, the analysis is centered at pre-service teachers' use of the word *parallelogram* to illustrate the differences in their discourse. The pre- and post-interview results show that the pre-service teachers' use of the word *parallelogram* changed from describing parallelograms as collections of unstructured quadrilaterals based on family appearances, to using the label *parallelograms* as collections of quadrilaterals sharing common descriptive narratives (i.e. definitions). The results also suggest that at van Hiele Level 3, some of the pre-service teachers have developed the concept of quadrilaterals with a hierarchy of classification, and some of them did not.

I now use Molly and Judi's discourse to discuss the scenarios of 1) changes in discourse when moving towards a higher van Hiele level of thinking, and 2) differences in discourse at the same van Hiele level thinking. In the case of Molly, her geometric discourse extended and the changes in her discourse from a lower level to a higher level of geometric thinking are detected by her use of the word *parallelogram*. For example, at a lower van Hiele level (Level 1), she stated, "this is a *parallelogram*" based on its family appearance and used the word *parallelogram* as a label to match polygons fitting this visual description, rather than the definition. While Molly was developing a concept of parallelograms and moving to a higher van Hiele level (Level 3), she used the word *parallelogram* as a collection of quadrilaterals (i.e. rectangles, squares, rhombuses, and parallelograms) sharing a common descriptive narrative: "a *parallelogram* is a quadrilateral with two pairs of parallel sides" (a common endorsed narrative used by the pre-service teachers in the study). Molly's geometric discourse was extended; however, she did not demonstrate a hierarchy of classification among parallelograms. In contrast, at the same van Hiele level, Judi demonstrated that parallelograms were connected with a hierarchy of classification when she used the word *parallelogram*. Judi and Molly represent a case of pre-service teachers whose thinking are at the same van Hiele Level (Level 3), but their use of the word *parallelogram* is different; one demonstrates an understanding of how parallelograms are connected with a hierarchy of classification, and the other one does not.

The interview results also identify the changes in pre-service teachers' geometric discourse while the test results suggest that there is no change in their van Hiele levels. For example, Sam's use of the word *parallelogram* changed from including all polygons having pairs of parallel sides without considering all the necessary conditions, to including only quadrilaterals having two pairs of parallel sides without a hierarchy of classification. This result reminds us that a single van Hiele level encompasses a range of complexity in using the word *parallelograms* and differences in discourse.

In addition, the changes in pre-service teachers' *routines* such as identifying routines, defining routines, verifying routines and substantiating routines suggest that these pre-services teachers have developed competence to use definitions to identify quadrilaterals, and some of them are able to construct informal and formal proofs. Briefly stated, the interview results show that pre-service teachers' routines changed from identifying polygons using visual recognition, to identifying them using endorsed narratives. In verifying their claims, some pre-service teachers' routines changed from recalling, measuring and/or constructing routines, to formulating proofs using mathematical propositions and axioms or using algebraic reasoning to verify claims in geometry.

For some pre-service teachers, their routines of verifying their claims are descriptions of processes of mathematical activity, whereas some pre-service teachers use algebraic reasoning. For example, Sam would verify that diagonals in a rectangle have equal measure by explaining, "They are the same because I measured it." The term "I measured it" reveals that her routine of verifying relies on comparing and checking measurements, and is a description of what *she did*. In a different scenario, Sam verified that two angles were congruent by asserting, "the angle can slide over to this position and create this angle, and the line can be rotated so that this angle now becomes this angle." The use of "angle" with "slide over" and "create," and the use of "line" with "be rotated" and "becomes," indicate that her routine of verifying the claim was a description of the process of what *the lines and angles did* through transformations, but such a description does not constitute a formal mathematical proof. In a different case, Ivy favored algebraic reasoning in her substantiation routines to justify her claim of "the diagonals bisect the angles of the square." For example, Ivy labeled the angles X and Y, and solved them algebraically, to find that they were 45 degrees each. Using this information, Ivy concluded that the diagonals bisect the angles [of the square]. Using algebra to help solve problems in geometry was quite common during the interviews.

The results of pre-service teachers' routines not only provide information on what they did as course of actions during the interviews, but also reveal their reasoning, problem solving strategies, and abilities to construct mathematical proofs.

5.2   What Can We Generalize and Why?

The van Hiele model describes geometric thinking in different stages; however, results of the study suggest that a single van Hiele level of thinking encompasses a range of complexity of reasoning and differences in discourse. The complexity of geometric thinking is seen through two of the most revealing characteristics of geometric discourse, *word use* and *routines*. In particular, word use provides information on how discursive *geometric objects* are perceived, whereas *routine of substantiation* describes how well defined (mathematical) practices are employed. I distinguish the term "geometric object(s)" which refers to all the mathematical objects involved in a particular geometric discourse (e.g., primitive objects or discursive objects), from the term "geometric figure(s)" which refers to all polygons. The discussions about pre-service teachers' substantiation at each van Hiele level focus on two types of substantiations: the object level and abstract level substantiations.

The object level substantiation emphasizes pre-service teachers' routines of substantiation, looking at descriptions of how geometric figures are being investigated. Describing static lines, angles and polygons as concrete entities as a way of substantiation, is an example of the object level of substantiation. With regard to definitions of different quadrilaterals, however, routines of substantiation depending on measurement routines to check the sides and angles of quadrilaterals, without thinking about how quadrilaterals are connected, are other examples of the object level of substantiation. Object level substantiation is a routine of substantiation, where students focus on the concreteness of geometric figures.

Abstract level substantiation emphasizes pre-service teachers' routines of substantiation using endorsed narratives to endorse new narratives such as use of mathematical definitions and axioms to construct mathematical proofs. During the interviews, pre-service teachers with an abstract level of substantiations also used object level substantiations to modify their justifications. For example, a student used the Angle-Side-Angle congruence criterion to construct a proof at an abstract level that opposite angles of a parallelogram are congruent, and could also justify informally why this congruence criterion works using rotations at an object level. In the following, I discuss characteristics of geometric discourse at van Hiele levels 1 to 4. See Appendix A for more detailed descriptions of geometric discourse at each van Hiele level.

5.2.1 The Level of Geometric Discourse

Geometric Discourse at Level 1
Level 1 visual-colloquial geometric discourse. Geometric figures are named based on their appearance. For geometric discourse at this level, the word use is passive. That is, the process of naming a polygon is an act of matching a picture of a polygon with its given name. When a student is asked for verification of why such polygons are called "rectangles", or why "opposite sides and angles of a parallelogram are equal", the course of actions include direct recognition, which is self-evident. Some students use their prior experiences to draw conclusions, but this course of action is known as rote memorization, such as "I learned it in school," or "I know it is a square." At this level, grouping quadrilaterals into different groups (i.e., rectangle, rhombus, parallelo-gram, square, etc.) is about putting them together by their names. Therefore, at *visual-colloquial* geometric discourse, *geometric objects* are collections of concrete, unstructured, discursive objects, and there is *no routine of substantiation*.

Geometric Discourse at Level 2
Level 2 visual-descriptive geometric discourse. Some properties of
geometric figures are identified, but they are not yet ordered. For
geometric discourse at this level, word use is routine driven, which
means that naming a polygon involves not just matching a polygon with
a name, but referring to it with a common descriptive narrative according
to some visual properties. For example, when a pre-service teacher is
asked for an explanation of why a polygon is called a "rectangle," or why
"opposite sides and angles of a parallelogram are equal," the courses of
actions include direct recognition, as well as counting the number of
sides, or measuring the sides and angles. Another possible response
could be, "It looks like it has four right angles," or "I measured and all the
angles are 90 degrees." At this level, grouping quadrilaterals into
different groups involves organizing them by their names and by some of
their visual properties. Therefore, at *visual-descriptive* geometric
discourse, *geometric objects* are collections of concrete, unstructured
discursive objects that might be placed into disjointed categories (i.e.,
they all have right angles, or parallel sides, etc.), and *routines of*
substantiations focus on checking and verifying partial visual properties
of geometric figures.

Geometric Discourse at Level 3
Level 3 *descriptive-theoretical geometric discourse.* Properties of
geometric figures are ordered, and they are deduced one from another.
Although at this level deduction is not understood, the definitions of
figures come into play. For geometric discourse at this level, word use is
still object-driven, as the naming of a polygon depends on its visual
properties, and a common descriptive narrative accompanying the name
of the figure (i.e., a definition of a quadrilateral). For example, when a
pre-service teacher is asked why a polygon is called a "rectangle," the
course of action is to check the defining conditions of the polygon by
counting the number of sides, and measuring and comparing the sides
or angles. Another possible response could be, "It is a rectangle
because it is a parallelogram, and it has four right angles." At this level,
when grouping quadrilaterals, a polygon can belong to multiple groups

at the same time by definitions. A concrete discursive object, such as a 4-sided polygon is labeled as a "square" in a previous geometric discourse level, becomes an abstract discursive object at this geometric discourse level, That is, a square is identified as a rectangle, a parallelogram, and a rhombus because it fits the definitions of parallelogram, rectangle and rhombus. Therefore, at *descriptive-theoretical geometric discourse, geometric objects* are collections of concrete discursive objects and they begin to connect with joint categories. In the case of quadrilaterals, all 4-sided polygons begin to fall into a hierarchy of classification. *Routines of substantiations* reveal a range of informal and formal deductive reasoning at an early stage.

Geometric Discourse at Level 4
Level 4 *deductive geometric discourse.* Students reason deductively. For geometric discourse at this level, word use is object- driven. That is, common descriptive narratives (i.e., definitions) guide the activity of naming of a polygon or a mathematical term (e.g., angle bisector, supplement angle, etc.). Grouping quadrilaterals into different groups means arranging them by definition with a hierarchy of classification. Routines of substantiations lead to the construction of new endorsed narratives. At this level, using definitions is more fluent in substantiation and in making connections among endorsed narratives (axioms, propositions, etc.) to construct new endorsed narratives. Therefore, at *deductive geometric discourse, geometric objects* are collections of abstract discursive objects, and the *routines of substantiation* consist of all the routines for substantiation in the previous geometric discourse, but the main activity of substantiation is to produce newly endorsed narratives, or commonly, to construct mathematical proofs.

　　The descriptions of geometric discourse at each van Hiele level indicate that the subsequent level of mathematical discourse is a meta-discourse of the former one. Viewing van Hiele levels of thinking as different levels of geometric discourse expands our understanding about geometric thinking and its development through the variability of students' geometric discourse at the same van Hiele level, as well as at two consecutive levels.

5.2.2   Differences in Discourse Within a van Hiele Level

Sam and Judi showed no change in van Hiele levels, but there were changes in their geometric discourse. Sam's test responses suggested that her thinking operated at Level 2 – Level 2 (descriptive level). However, there were changes in her geometric discourse. For example, Sam's use of the word "parallelogram" changed from *any* polygon having pairs of parallel sides, in using a definition of parallelogram with only a necessary condition, to use the word with both necessary and sufficient conditions. The findings show that her thinking fit more towards the descriptions of geometric discourse at Level 3 at the end of the semester. Figures 5.1 and 5.2 illustrate the characteristics of Sam's geometric discourse in the pre- and post-interview, respectively.

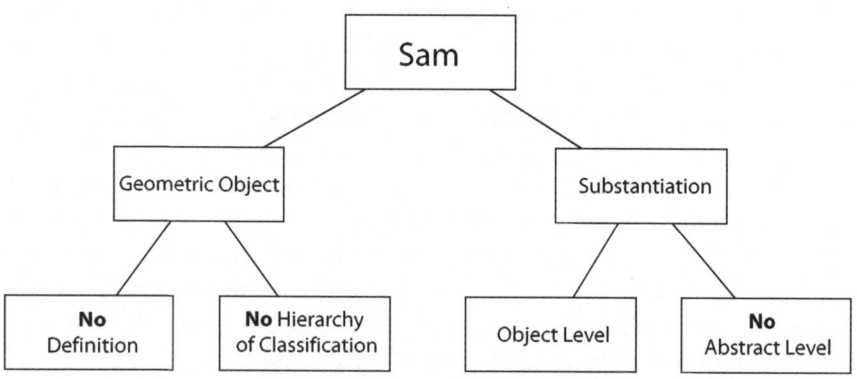

Figure 5.1. Characteristics of Sam's geometric discourse at Level 2 in the pre-interview.

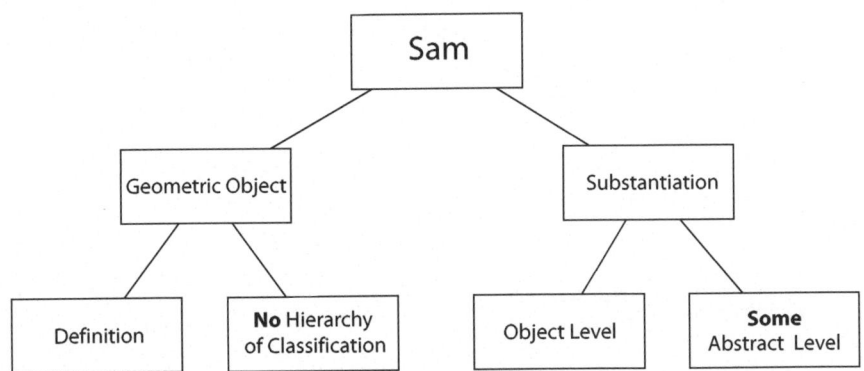

Figure 5.2. Characteristics of Sam's geometric discourse at Level 2 in the post-interview.

Figures 5.1 and 5.2 show two main changes in Sam's geometric discourse, a change in word use (Geometric Object) and a change in reasoning (Substantiation). Sam developed competence in using definitions to identify and group polygons but did not group them in a hierarchy of classification. She demonstrated some informal deductive reasoning at an abstract level as her geometric thinking moved towards Level 3. This observation does not contradict the findings from Sam's van Hiele geometry test results, but adds more information to treat the development of her thinking more dynamically. Sam represents a case when a student's geometric thinking develops continuously within Level 2 and in transition between Level 2 and Level 3.

Judi's geometric discourse presents another case of such continuity, but within a different van Hiele level. Her van Hiele test responses suggested that her thinking operated at Level 3 – Level 3 (deductive). Judi came in with the ability to identify and group polygons using definitions, and then began to reason more abstractly by constructing mathematical proofs using definitions and axioms. Figures 5.3 and 5.4 illustrate characteristics of Judi's geometric discourse at Level 3, in the pre-interview and the post-interview, respectively.

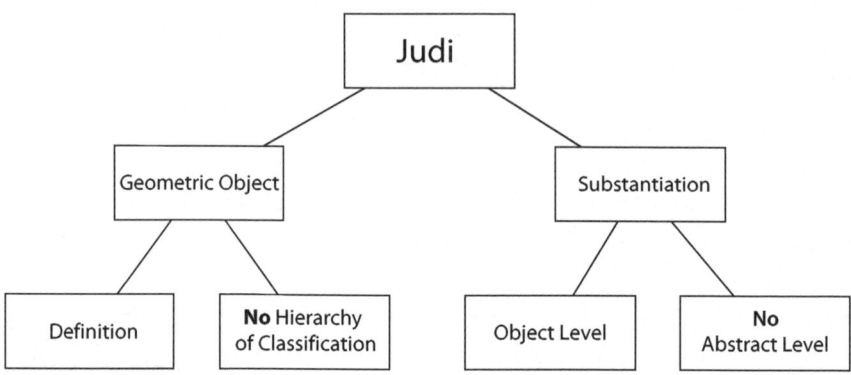

Figure 5.3. Characteristics of Judi's geometric discourse at Level 3 in the pre-interview.

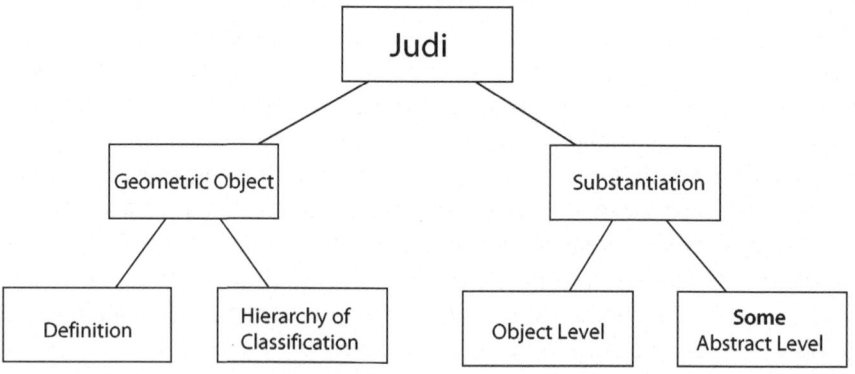

Figure 5.4. Characteristics of Judi's geometric discourse at Level 3 in the post-interview.

Judi was familiar with the use of definitions of quadrilaterals, and later she developed more competence in using definitions and demonstrated that these quadrilaterals were connected with a hierarchy of classification. Changes in Judi's geometric discourse within Level 3 were also evident as Judi used informal deductive reasoning as her routine of substantiation to justify her claims.

5.2.3 Differences in Discourse among Two Consecutive Levels

Kevin was one of two students in the study who reached Level 4 at the end of the semester. Similar to other students (e.g., Judi), at Level 3, Kevin demonstrated an understanding of using definitions, but when grouping quadrilaterals he did not show how quadrilaterals were connected with a hierarchy and performed object level substantiations as illustrated in Figure 5.5.

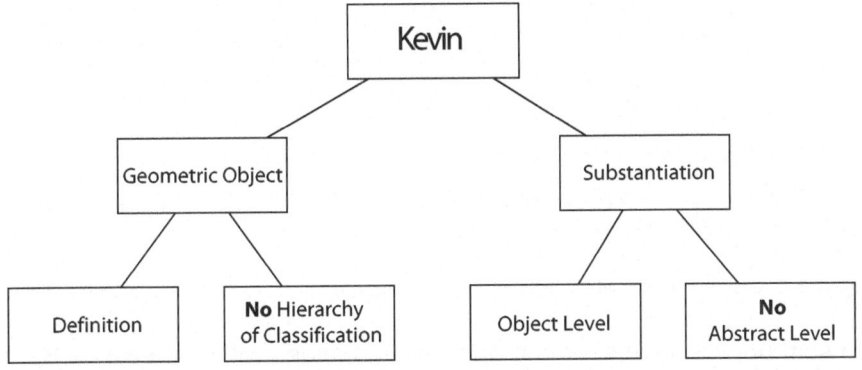

Figure 5.5. Characteristics of Kevin's geometric discourse at Level 3 in the pre-interview.

In contrast, he was able to draw connections among quadrilaterals and to use propositions and axioms to construct mathematical proofs. Figure 5.6 presents a main characteristic of a Level 4 (abstract) discourse that is absent in Level 3: abstract level of substantiations. At this level, Kevin showed familiarity with using definitions and axioms to construct proofs and with using algebraic symbols to write a formal mathematical proof. However, at Level 4 students are able to apply inductive reasoning in an unfamiliar situation, and to connect the knowledge they learned. In Kevin's case, he was able to apply his knowledge of quadrilaterals to construct mathematical proofs in a familiar situation (e.g., to prove opposite angles or sides are congruent using congruent criterion), having carried out similar proofs in his geometry class. When Kevin was asked to complete the task that was new to him, he did not know how to use the same axioms in a new situation. The findings suggest that Kevin was at the beginning stage of Level 4 thinking, starting to gain the skills and languages needed for mathematical proofs, but needing more practice to move forward to an advanced abstract level.

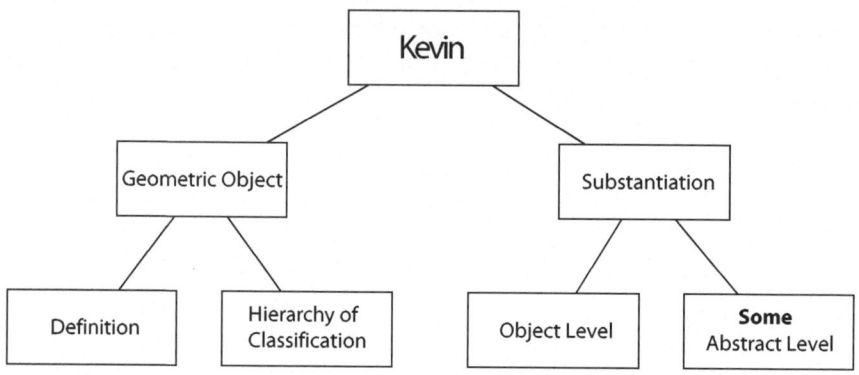

Figure 5.6. Characteristics of Kevin's geometric discourse at Level 4 in the post-interview.

To sum up, we have seen the discursive changes within a level (in Sam and Judi's case), as well as at two consecutive levels (in Kevin's case). Using Sfard's communicational approach to investigating geometric thinking allowed for the details needed to have a better understanding of what students said about geometric figures and what they did as a course of actions in a discourse specific context. The results of the study also suggest that a deeper investigation in deed, focusing on discursive differences in students' geometric thinking to consider the possibilities of *other* characteristics at each van Hiele level.

5.2.4 *Characteristics Should Be Considered at the van Hiele Levels*

The van Hieles also argue that a learning process is also a process of learning a new language, because "each level has its own linguistic symbols" (van Hiele, 1959/1985, p. 4). The van Hiele levels reveal the importance of language use, and language is a critical factor in the movement through the levels. van Hiele (1986) provides an explanation of the language use at each level. For example, at the Basic level there is a *language*, but the use of this *language* is limited to the indication of configurations that have been made clear based on observation. At Level 1, students need to develop the *language* that belongs to the descriptive level. At Level 2, the *language* has a much more abstract character then the descriptive level, and reasoning about logical relations between theorems begins at this level. At Level 3, students use the *language* of proof (pp. 43-53). However, the word "language" is not clearly defined in the way it is broadly used. One would consider "language" in the comparisons of informal language versus formal language, whereas others would refer to it as the different use of mathematical vocabulary at different van Hiele levels.

The van Hieles wished to note language differences and different linguistic symbols at each level, but never provided explicit descriptions. For instance, the van Hiele descriptions of the levels focus largely on how a student reasons about geometric figures in a language, in response to what is a rectangle versus what is not a rectangle, and applying a definition. What is missed or not clear emphasizes the meaning of a mathematical term when used by a student. When each van Hiele level is considered as its own geometric discourse with characteristic of mathematical terms and their use (word use), it reveals facts concerning how a concept is formed through the use of a mathematical word (e.g., parallelogram, rectangle, square, etc.). In this study, word use provides significant information about how a concept of a geometric figure called "parallelogram" is developed at different van Hiele levels among different pre-service teachers. Moreover, a careful analysis of pre-service teachers' mathematical word use in geometric discourse also sheds a light on how words are used and whether the words are used correctly for the sake of communication.

Discursive routines do not determine students' actions, but as negotiated conventions only constrain what they can reasonably say or do in a given situation. However, discursive routines offer valuable information about a student's course of actions to make conjectures and justifications as a pattern of geometric discourse. The analysis of these actions brings details of students' routines of identifying, defining, and justifying when working on a task about geometric figures and their properties, where the roles of definitions at the first three van Hiele levels are brought to light. Discursive routines are also associated with students' creativity when they apply routines in non-routine ways, such as applying familiar routines in an unfamiliar discursive context. For example, in this study, some pre-service teachers used algebraic reasoning (familiar routines) to construct geometric proofs (unfamiliar discursive context), without using geometric axioms.

The discourse approach has offered a different lens to better understand geometric thinking and its development. Looking at what we do know about mathematical thinking, language, and discourse practices in the classroom, we are led to ask what future studies are needed in this important area of research.

5.3   What Can Be Asked and Why?

Several studies using van Hiele levels to determine students' levels of geometric thinking across different topics indicate that students may not be working at the same level on all concepts (e.g., Burger & Shaughnessy, 1986; Mayberry, 1983). For example, Mayberry (1983) assessed nineteen undergraduate pre-service teachers' levels of thinking using seven geometry concepts: squares, right triangles, isosceles triangles, circles, parallel lines, similarity, and congruence. The study found that "the determination of the success criterion for a given topic and level was rather subjective" (p. 68). This conclusion can be understood to indicate that pre-service teachers' geometric thinking was at different van Hiele levels for different concepts. However, the study did not explore further in what ways they are different. For instance, one might suspect that a more difficult concept such as similarity would require a higher van Hiele level of thinking than the classification of a quadrilateral. In Burger and Shaughnessy's project (1986), interview tasks consisted of drawing, sorting, identifying, and defining geometric shapes such as triangles and quadrilaterals. With regard to different tasks, some students operated at different levels of thinking. For example, one student was reasoning at Level 3 (Abstraction) on the sorting task, but was assigned to Level 4 (Deduction) on the identifying and defining tasks because he was able to conjecture and attempt to verify his conjecture by means of formal proof (p.42). Towards that end, this study took a different direction in examining students (pre-service teachers)' geometric thinking through their geometric discourse. The results of these examinations revealed small fractions of the richness of human thinking, while helping to add a little more knowledge to what we know *so far* about geometric thinking and its development. However, the results of the study show that a single assignment of a van Hiele level does not encompass a range of complexity of human thinking and reasoning. Thus, deeper investigations of students' mathematical thinking and its development with multiple lenses are warranted.

5.3.1 What Do We Learn from Students' Thinking?

Some researchers suspected the existence of a level (Level 0) prior to the Base Level (Level 1). As a result of the study, it shows that students can reason at a higher van Hiele level, but their lack of knowledge in geometry, or simply forgetting what they learned in geometry, has kept them from giving correct answers. In Ivy's case, the geometric pre-test placed her at Level 0. However, Ivy's geometric discourse revealed that she was able to group quadrilaterals by their names, but did not know the differences between a rhombus and a square, or the differences between a parallelogram and a rectangle; Ivy's geometric discourse fit more to the descriptions at Level 1 than Level 0.

It was quite common that a student could not identify a "trapezoid," or a "rhombus," because they did not learn these names, or forgot the names. So, if we consider the existence of Level 0 (a pre-level to Level 1), then it is likely that we include the possibility of the kind of reasoning students perform in a domain of knowledge that they have not yet explored. This observation led us to consider, "What does the van Hiele model of thinking serve to assess?" and "What do we wish to find out using the van Hiele model of thinking?"

Next, consider this scenario of students being prompted to show that two opposite angles were congruent in a parallelogram. First, they were visually convinced that the two angles were the same, but further verification was required. One student responded that the two angles were congruent because she used a protractor to measure the angles and they had the same measurement. This course of action is typical in geometric discourse at Level 2, where a student's reasoning depends on checking and verifying the conditions for being congruent. From this response it could be argued that this student has mastered knowing what are "opposite angles" in a parallelogram, but needs to explore in a concrete way what we call "congruent," a property of opposite angles. For the same task, another student described a sequence of transformations where a rotation was followed by a translation, to show that the two angles were the same. She stated that she could rotate one angle, moved the angle to match the other one, and was sure that the two angles would match exactly. This course of action is typical in geometric discourse at Level 3, where a student is familiar with the term "congruent" and tries to concretely explore whether opposite angles are congruent. From the two responses, I would argue that students need to explore the properties of parallelograms through hands-on activities before they reach the conclusion that "*all* opposite angles are congruent in *any* parallelograms"; then inductive reasoning starts to make sense.

The van Hiele levels are sequential, in that students pass through the levels in the same order, although varying at different rates and it is not possible to skip levels. Some observations from the study suggest that it would be beneficial for a student to understand "opposite angles" in the case of a parallelogram and the meaning of "congruent" first, then move on to explore the properties and relations regarding opposite angles in a parallelogram. Perhaps, based on these concrete experiences, the student would begin to develop some abstract thinking such as inductive reasoning. Thus, the object level of substantiations clouds important points van Hiele made in that students must explore domains before describing them, and that elaborate descriptions of concrete properties and relations must be made before abstract relations are explored.

Another observation from the study is the challenge students face in developing abstract relations, because the abstract relations in geometry may never be fully understood by some students. It takes time for students to get used to new mathematical terms, as well as to digest the hands-on activities relating to a particular property, before they can generalize it. When students are introduced to more advanced thinking in deductive reasoning, some mimic the proofs without fully understanding them. When we rush to the stage of constructing proofs that a student is not ready for, it creates obstacles. It is important to give students enough opportunities to explore a sequence of activities at a level built on other activities at a previous level before abstract relations are explored.

Lastly, what is the development of Level 3 geometric thinking? The results of the study show that students being placed in van Hiele Level 3 demonstrated the competence in using definitions to identify quadrilaterals. However, results also indicate that geometric discourse at Level 3 varied from person to person and varied in the same person at different times of the semester. Two main variations of the discourse at Level 3 are 1) how profound students use definitions (geometric object) and 2) the way they reason about the geometric figures (substantiation). Figure 5.7 illustrates these two variations in Level 3.

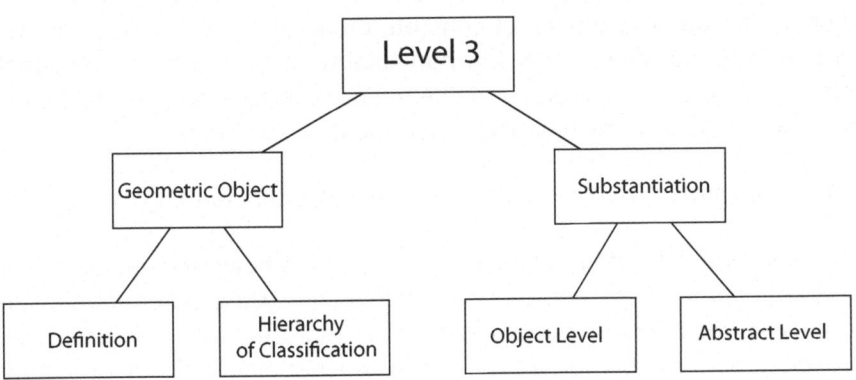

Figure 5.7. Characteristics of Level 3 geometric discourse.

Figure 5.7 highlights possible variations of geometric discourse at Level 3. Having a geometric discourse at Level 3 indicates that a student has developed competence in applying definitions in their identifying and justifying routines. In this study, such a student may or may not make connections among the quadrilaterals where a hierarchy of classification is presented and doing so is dependent on how profoundly the student understands the definitions and uses them adequately.

Molly's geometric discourse shows that a student at Level 3 could have competence in using definitions to identify quadrilaterals, but still needs to develop other skills needed at this level. In contrast to Molly's geometric discourse at the same level, Sam and Kevin represented a group of students who could use definitions fluently as well as reason at the object level. Ivy and Judi represented a group of students who were more advanced at Level 3, when they used definitions to show a hierarchy of classification among quadrilaterals; based on their experiences of reasoning at object level, they also tried to substantiate their conclusions using some abstract relations.

These results suggest a variety of geometric discourses at Level 3, and help us to learn that van Hiele Level 3 thinking is more complicated than previously thought. At this level, students need to be familiar with and feel comfortable using the definitions fluently *and* at the same time, they are also developing informal reasoning by describing what they observe through exploration of concrete properties of geometric figures, so that they can see or feel the particularity of figures before abstract relations take place. All these mathematical activities become students' prior experiences in the development of abstract thinking.

5.3.2 *Classroom Discourse Practices and Student Learning*

The empirical data of the study offers a dialogue between a student and a researcher through one-on-one interactions, using carefully chosen tasks with well-designed questions to prompt student's thinking. This procedure is not too far from what a teacher might do in preparing instructional materials. For example, teachers' often make efforts to actively envision how students might mathematically approach instructional tasks they will work on. As described in Smith & Stein's framework (2011), "anticipating students' responses involves developing considered expectations about how students might mathematically interpret a problem, the arrays of strategies and how these interpretations and strategies might related to mathematical concepts" (p.8).

The concerns about "communication," "language," and "discourse" in the mathematics classrooms are not new. Fifty years ago, the same concerns motivated the van Hieles to develop the model to better understand geometric thinking and how it is communicated in the classrooms, and to improve teaching and learning. The results of the study add information on the need to be explicit about the mathematical terms and their use in the classrooms. For example, when a student mentions the word "parallelogram," she/he says nothing unless she makes explicit what she means. We have seen that the word "parallelogram" could mean quite different things to pre-service teachers in the study. Some pre-service teachers thought parallelograms were four sided figures having two pairs of parallel sides with two sides longer, whereas some of them thought parallelograms were tilted rectangles and squares. A few of them thought that a parallelogram was a polygon with pairs of parallel sides, and then of course, hexagons and octagons were parallelograms. Teacher educators need to be very cautious and communicate well with pre-service teachers about the way they use mathematical terms, in order to make classroom discussions more productive. Therefore, questions such as, "Can you say more about what you mean by parallelogram?" "Will you give me an example of a parallelogram?" "Can you draw a different parallelogram?" and "Why do you think these are different parallelograms?" can serve as formative assessment tools to help elicit student thinking about the concept, and these questions could also serve as tools for conducting productive classroom discussions.

Many recent calls for improvements in instructional practices have focused on promoting deep and robust mathematical reasoning. Teachers are facing challenges in making this vision a reality in their classroom practices. There is a need for building a professional knowledge base for teachers and educators identifying effective classrooms practices to support students' meaningful learning in mathematics. In addition, there is a need for teacher professional development in mathematics instruction to create conversations among teachers to make the nature of mathematical language practices themselves a point of discussion about classroom interactions and discourse, and involving teachers in reflection, discussion and continuous interaction with colleges about issues of language in teaching and learning of mathematics.

5.4 Closing

This study offers an important contribution to the body of theoretical and empirical work in discourse in the very important, and often neglected, area of geometry, while also fueling improvement of teacher preparation and education, and classroom discourse practices. In particular, this study pilots an analytic method for investigating students' geometric thinking using a discursive lens, looking in particular at pre-service elementary teachers' geometric discourse in the context of quadrilaterals.

After some reflection, it seems to me that what is presented in the book (the empirical part of the work) is an example of the phenomenon regarding student geometric thinking and their understanding of mathematics. The discourse approach provided a tool to analyze students' understanding of mathematics with a detailed consideration of "*who* understands *what* and *when*" (Thurston, 1994). The empirical data also led to producing a framework, *the Development of Geometric Discourse*, to describe students' (pre-service teachers) geometric thinking. The advantages of using Sfard's communicational approach to mathematical discourse diverge from my experiences as a researcher and a teacher educator, and from my observations of teachers in their mathematics classrooms. From a research perspective, the study illustrates the usefulness of the discursive lens for highlighting the opportunities for rich description at each van Hiele level through discursive terms such as *word use, routines, endorsed narratives,* and *visual mediators*. Thus, it enhances our knowledge about student geometric thinking, their knowledge in geometry, and the ways they communicate their understanding about geometric shapes. From a teaching perspective, there is also an advantage of using a communicational approach to improve instructions—that is, to improve classroom discourse practices. Across her work, Sfard discusses a communicational approach to cognition around mathematical discourse focused on learning. However, I am convinced that the communicational approach has implications for teaching as well. For instance, the communicational approach to learning mathematics suggests that learning is a special social interaction and knowing of mathematics is the ability to participate in mathematical discourse. That provides a theoretical ground for teachers to conduct dialogues and productive discussions in the classrooms to engage students thinking and speaking mathematically through social interactions.

While this book is aimed at how the discourse approach to geometric thoughts could add more knowledge to student mathematical thinking, the empirical data presented in the book also provides evidence on how mathematics instruction might have contributed to help interpret the changes between the pre-test and post-test, and changes in pre-service teachers' (students) geometric discourse. For example, In attempting to peel back layers of assumptions, it is important to discuss how the mathematics instruction might have led thirty pre-service teachers (about 47%) to van Hiele Level 3 thinking at the post-test, and at least sixteen pre-service teachers moved up at least one van Hiele level between the pre-test and post-test. These results suggest that the mathematics instructions on geometry had an impact on these pre-service teachers' learning of the concepts of quadrilaterals. This reminds us of the importance of academic instructions in students' learning and in the development of mathematical concepts. More importantly, the end result of categorizing students' thinking or assigning a level of geometric discourse (or a van Hielel level) should not be used as "a label," but to advance our knowledge about learning, and therefore improve our mathematics instruction.

The methodological contribution of the study is to revisit the van Hiele model of geometric thinking with a new framework, take on a new perspective, and develop a new theory about mathematical thinking and discourse. For example, conducting a careful analysis on pre-service teachers' *word use* at each van Hiele level shed a light on their understanding of the concepts of quadrilaterals through the use of the word "parallelogram," and it captured the salient relationship between a word and a mathematical concept. Drawing on a Vygotsky's (1987) view, "the development of scientific concepts begins with verbal definition, and the development of concepts and the development of word meanings are one and the same process" (p.168). Thus, a deeper investigation of connections between students' use of mathematical words and the development of mathematical concepts is warranted to advance our understanding of concept development. Although the setting of the study was one-on-one interviews, it provided opportunities for dialogues between the researcher and pre-service teachers. In this setting, pre-service teachers could draw pictures and diagrams, and use body language and gestures to communicate their thinking. The researcher could prompt different questions to elicit pre-service teachers' thinking. However, in a classroom of twenty students, conducting dialogues and discussions in the classroom and eliciting students thinking is a different and challenging task. Thus, future research in the area of discourse practices in mathematics classrooms is needed to expand our understanding about a communicational approach to teaching. The study compared pre-service teachers' geometric discourse from two ends, pre-interview and post-interview, to identify possible changes in their learning. Thus, future research is needed to focus on more than brief interactional dialogue fragments and examine how mathematical discourse evolves over a long period of time.

It is my sincere hope that this book helps to bring opportunities for conversations among researchers, curriculum developers, teacher educators, and teachers on the learning and teaching of geometry. Such conversations would address the complexity of students' geometric thinking and the development of geometric concepts, and the challenges of discourse practice in mathematics classrooms. As a start, we need to embrace the multi-semiotic nature of mathematical activity and multi-faceted views of mathematical discourse.

Appendices

Appendix A: Translation of van Hiele Levels of Geometric Discourse

Description of Level 1		The Visual-Colloquial Geometric Discourse	
Van Hiele level 1 (Visual): "Figures are judged by their appearance. A child recognizes a rectangle by its form and a rectangle seems different to him than a square. At [this] level, a child does not recognize a parallelogram in the shape of rhombus" (van Hiele, 1959/1985, p. 62)	*Selected Van Hiele Quotes* 1. "Figures are judged according to their appearance." 2. "A child recognizes a rectangle by its form, shape 3. and the rectangle seems different to him from a square." 4. "When one has shown to a child of six, a six year old child, what a rhombus is, what a rectangle is, what a square is, what a parallelogram is, he is able to produce those figures without error on a geoboard of Gattegno, even in difficult situations."	Word Use	1. The *names* of geometric objects are judged with their appearances: parallelogram, rectangle, square, etc. 2. The *use of verbs* is connected to the concrete objects: see, looks like, it is, etc.
		Routines	1. *Direct recognition*: "what one sees about geometric objects" For example, "this is a rhombus," "this is a parallelogram" "parallelogram is not a rhombus. The rhombus appears… as something quite different." 2. *The routine procedure is a perceptual experiences and it is self-evident.* For example, when asked for substantiation of why "This is a rhombus", one would say, "because it looks like one"

| | 5. "a child does not recognize a parallelogram in a rhombus."

6. "the rhombus is not a parallelogram. The rhombus appears … as something quite different." | Endorsed Narrative | *Some examples of endorsed narratives:*
1. "this one (a square) looks different than this one (a rectangle)."
2. "a rhombus is not a parallelogram because a parallelogram has two sides longer than the other two." |
| | 7. "when one says that one calls a quadrilateral whose four sides are equal a rhombus, this statement will not be enough to convince the beginning student [from which I deduce that this is his level 0] that the parallelograms which he calls squares are part of the set of rhombuses."

8. (on a question involving recognition of a titled square as a square) "basic level, because you can see it." | Visual Mediators | *Visible objects* that are operated upon as a part of the process of direct recognition:
1. 2-D geometric shapes (e.g., triangles, quadrilaterals, etc.)
2. Angles (e.g., angles look like right angles, angles look like greater, or smaller than a right angles, etc.)
3. Lines (e.g., two lines look parallel, two line look perpendicular, etc.)
4. The physical orientations of a geometric figure.
For example: two identical squares as, one would say, the one on the left is a square, and the one on the right is a rhombus. |

Description of Level 2		The Visual-Descriptive Geometric Discourse	
van Hiele level 2 (Descriptive): "Figures are bearers of their properties. That a figure is a rectangle means that it has four right angles, diagonals are equal, and opposite sides equal. Figures are recognized by their properties. At this level properties are not yet ordered, so that a square is not necessarily identified as being a rectangle" (van Hiele, 1959/1985, p. 62)	*Selected Van Hiele Quotes*: 1. "He is able to associate the name 'isosceles triangle' with s specific triangle, knowing that two of its sides are equal, and draw the subsequent that the two corresponding angles are equal." 2. "… a pupil who knows the properties of the rhombus and can name them, will also have a basic understanding of the isosceles triangle = semirhombus." 3. "That a figure is a rectangle signifies that it has four right angles, it is a rectangle, even if the figure is not traced very carefully." 4. "The figures are identified by their properties. (e.g.) If one is told that the figure traced on the blackboard	Word Use	1. The *names* of geometric objects are associated with their properties. For example, the word "isosceles triangle" signifies not any triangle but a special triangle, which has two sides that are equal, and because of that it also signifies the two corresponding angles are equal. 2. The use of words such as "diagonal" "transversal" "perpendicular" "bisect" 3. The use of verbs is *personal*. For example, "I rotated this figure…" or "I moved it to…"
		Routines	1. The routine procedures include *substantiation*[1] and *recall*[2], however the *construction*[3] of writing mathematical proofs is not yet developed. For example, a student recognizes an object is a "rectangle," and also explains that "an object is a rectangle because it has four right angles" after *checking* the measurements of

---

[1] *Substantiation*, the action that helps one to decide whether to endorse previously constructed narratives.

[2] *Recall*, the process one performs to be able to summon a narrative that was endorsed in the past.

[3] *Construction* is a discursive process resulting in new endorsable narratives.

	possesses four right angles, it is a rectangle, even if the figure is not traced very carefully."		the angles of the object.
	5. "The properties are not yet organized in such a way that a square is identified as being a rectangle."	Endorsed Narrative	*Some examples of endorsed narratives*: 1. "Squares are not rectangles because squares have all sides equal, but rectangles do not." 2. "Isosceles triangles have two base angles that are equal." 3. "Diagonals of a rectangle are equal." 4. "Diagonals of a parallelogram bisect each other."
	6. "The child learns to see the rhombus s an equilateral quadrangle with identical opposed angles and interperpendicular diagonals that bisect both each other and the angles."	Visual Mediators	*Visible objects* that are operated upon as a part of the process of direct recognition: 1. 2-D geometric shapes (e.g., triangles, quadrilaterals, etc.) 2. Objects are identified by their properties. For example, if one is told that the figure in the picture has four equal sides, then this figure is a rhombus, even if the figure is not drawn very carefully.

Description of Level 3		The Descriptive-Theoretical Geometric Discourse	
van Hiele level 3 (Theoretical): "Properties are ordered. They are deduced one from another: one property precedes or follows another property. At this level the students do not understand the intrinsic meaning of deduction. The square is recognized as being a rectangle because at this level, definitions of figures come into play" (van Hiele, 1959/1985, p. 62)	*Selected Van Hiele Quotes*: 1. "Pupils … can understand what is meant by 'proof' in geometry. They have arrived at the second level of thinking." 2. "He can manipulate the interrelatedness of the characteristics of geometric patterns." 3. "e.g., if on the strength of general congruence theorem, he is able to deduce the equality of angles or linear segments of specific figures." 4. "The properties are ordered [lit. 'ordonnent']. They are deduced from each other: one property precedes or follows another property."	Word Use	1. The *names* "parallelogram," and "rectangle" signify the *realizations* of geometric figures based on the *narratives* of these figures. For example, the word "rectangle" signifies a parallelogram with four right angles" based on the definition of rectangle." And "a square is recognized as being a rectangles by definition." 2. The use of words "prove", "imply/implies," "equivalence/equivalent"
		Routines	1. The routine procedures involves *substantiation* and *recall* as in the Level 2. 2. The construction of informal proofs. For example, to explain, "opposite angles in a parallelogram are equal." one would say, "if the angle has been rotated 180°, they will match exactly, so opposite angles are equal."
		Endorsed Narrative	*Some examples of endorsed narratives*: 1. "A rhombus is a parallelogram whose diagonals bisect each other perpendicularly" 2. "All equilateral triangles are isosceles triangles." 3. "A parallelogram has two pairs of parallel sides,

	5. "The intrinsic significance of deduction is not understood by the student."		this implies that two adjacent angles add up to 180°"
		Visual Mediators	*Visible objects* that are operated upon as a part of the process of direct recognition: 1. 2-D geometric figures such as triangles, squares, rectangles, other quadrilaterals, etc. 2. Some characteristics of a figure such as a pair of parallel sides of a quadrilateral, or the right angle of a triangle and corresponding *symbols*. 3. "Be able to deduce the equality of angles from parallel lines." For example, the alternate interior angles are recognized as part of a Z-form, interior angles on the same side of the intersecting line are recognized as part of a U-form, and corresponding angles are cognized as part of a F-form." 4. Be able to deduce the equality of vertical angles by recognition of an X-form.
	6. "The square is recognized as being a rectangle because at this level definitions of figures come into play." 7. "the child… [will] recognize the rhombus by means of certain of its properties,… because , e.g., it is a quadrangle whose diagonals bisect each other perpendicularly.		

Description of Level 4		The Deductive Geometric Discourse	
van Hiele level 4 (Deduction): " Thinking is concerned with the meaning of deduction, with the converse of a theorem, with axioms, with necessary and sufficient conditions" (van Hiele, 1959/1985, p. 62)	1. "He will reach the third level of thinking when he starts manipulating the intrinsic characteristics of relations. For example: if he can distinguish between a proposition and its reverse" [sic. Meaning our converse] 2. We can start studying a deductive system of propositions, i.e., the way in which the interdependency of relations is affected. Definitions and propositions now come within the pupil's intellectual horizon." 3. "Parallelism of the lines implies equality of the corresponding angles and vice versa." 4. "The pupil will be able, e.g., to distinguish between a proposition and its converse."	Word Use	1. The *names* "parallelogram," "rectangle" signifies the *realizations* of geometric figures based on the *endorsed narratives* of these figures. The endorsed narratives include definitions of geometric figures, axioms and theorems that are related to these geometric figures. For example, the word "rectangle" signifies the following: - "a parallelogram with four right angles based on the *definition of rectangle.*" - *the property of the rectangle*, "the diagonals of a rectangles are equal" - *the axiom related to the prove of the property*, "triangle congruence criterions." 2. The use of words "prove," "imply/implies," "equivalence/equivalent."
		Routines	1. The routine procedures involve *substantiation* and *recall* and *construction*. 2. The construction of formal proof. For example, to explain that "a parallelogram has all opposite sides equal," one would provide a formal proof: - First draw a diagonal which divides the parallelogram into two triangles. - Use Side-Side-Side criterion for congruence to prove that these two triangles are congruent.

	5. "it (is) … possible to develop an axiomatic system of geometry." 6. "The mind is occupied with the significance of deduction, of the converse of a theorem, of an axiom, of the conditions necessary and sufficient."	- Corresponding sides in the two triangles are equal 3. The use of mathematical symbols. For example, use mathematical notation such as "$\triangle ABC \cong \triangle ADC$" instead of "triangle ABC is congruent to triangle ADC"; Use "$\angle$" to indicate "angle", etc.
	Endorsed Narrative	*Some examples of endorsed narratives*: 1. Mathematical proofs (*Written*). 2. " to show the diagonal are perpendicular bisectors, you need to proof that two angles are equal and they add up to 180°, that will give 90° angles (perpendicular)." And "you also need to prove these two triangles are congruent so that all the sides are equal (bisect each other)." (*Verbal*)
	Visual Mediators	*Visual objects and mathematical symbols* 1. 2-D geometric figures such as triangles, squares, rectangles, other quadrilaterals, etc. 2. Symbols that represent parallel line (//), angles ($\angle$), equivalence ($\cong$), etc.

Description of Level 5		The Abstract Geometric Discourse	
van Hiele level 5 (Abstraction): "Figures are defined only by symbols bound by relations. [these] symbols belongs to a relational system which cannot be axiomatized because it cannot have direct liaison with logic" (van Hiele, 1959/1985, p. 64)	*Selected Van Hiele Quotes*: 1. "A comparative study of the various deductive systems within the field of geometric relations is … reserved for those, who have reached the fourth level…" 2. "the axiomatic themselves belong to the fourth level." 3. "one doesn't ask such questions as: what are the points, lines, surfaces, etc.? …figures are defined only by symbols connected by relationships. To find the specific meaning of the symbols, one must turn to lower levels where the specific meaning of these symbols can be seen."	Word Use	1. The *names* "rectangle" signifies the *realizations* of a geometric figure in both Euclidean and non-Euclidean geometry. 2. Geometric figures are signified only by symbols and connected by relationships. 3. The use of words in logic. For example, the "if P, then Q" statement.
		Routines	The routine procedure is considered as "creative"
		Endorsed Narrative	*Some endorsed narratives*: 1. "Squares are parallelograms with four right angles and four equal sides in Euclidean geometry, but in Taxicab geometry, a square represents a circle by definition."
		Visual Mediators	*The visual objects are mathematical symbols and artifacts* used in the domain of Euclidean and non-Euclidean geometry.

Appendix B: Interview Tasks

Task One

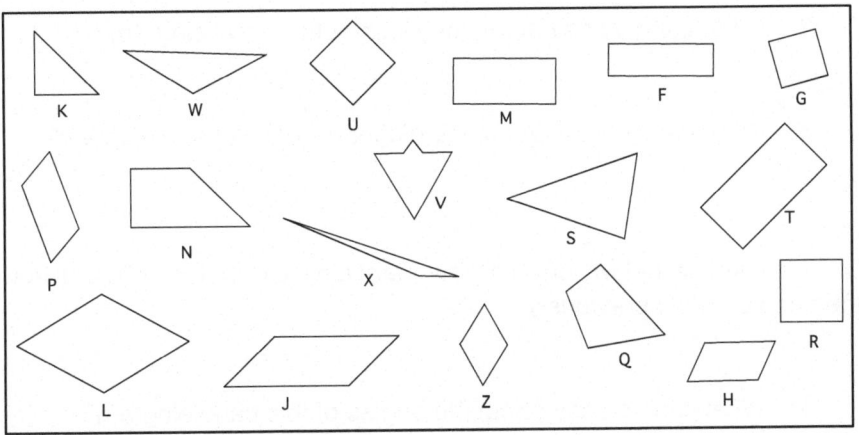

Figure Appendix B.1. Task 1: Sorting geometric figures.

Task Two

A. Draw a *parallelogram* in the space below.

    1.   What can you say about the angles of this parallelogram?

    2.   What can you say about the sides of this parallelogram?

    3.   What can you say about the diagonals of this parallelogram?

B. In the space below, draw a new parallelogram that is different from the one you drew previously.

    4.   What can you say about the angles of this parallelogram?

    5.   What can you say about the sides of this parallelogram?

    6.   What can you say about the diagonals of this parallelogram?

Appendix C: Interview Protocols

Before beginning the interview, provide the student with the following materials: Pencils, ruler, protractor, blank sheets of paper
Turn on both video cameras.

Task One

Present Task One and turn the page to face the student.

1. Say: These are geometric shapes. Sort these shapes into groups. You can sort them any way you want. Write down your answers at the bottom of the task, and make notes about why you group them in such a way. Let me know when you are finished.

While the student is working on the task, check the positions of the cameras and see if they are recording appropriately. Monitor the student while she/he is working on the task, and make notes to prepare possible questions.

After the student has finished the task, turn on the audiotape.

2. Ask: Can you describe each group to me?
After the student has finished describing her/his results, ask one of the following:

If the student sorts the shapes as all rectangles together, all triangles together, all squares together, etc, then
- Ask: Can you find another way to sort these shapes into groups? Try it.
- Ask: Why?

If the student sorts the shapes as all triangles together, all quadrilaterals together, etc., then
- Ask: Can you sort these shapes into subgroups? Try it.
- Ask: Why?

If the student says that he/she doesn't know any other way to sort the shapes, then
- Ask: Can "this" (e.g., a rectangle, or a parallelogram) and "this" (e.g, a rhombus, or a trapezoid) go together?
- Ask: Why, or why not?

3. Ask: What is a parallelogram?

After the student has answered the questions verbally, then give the student a piece of blank paper, and Say: write it down. Do the same for the following questions.

4. Ask: What is a rectangle?
5. Ask: What is a square?
6. Ask: What is a rhombus?
7. Ask: What is a trapezoid?
8. Ask: What is an isosceles triangle?

Turn off the cameras and audio recorder. Remind the student to write the date and his/her name on all the worksheets.

Say: I will collect all your worksheets.

Put all Task One materials away, give the student three minutes break and get ready for Task Two.

Task Two
Turn on both video cameras and audio recorder.

Present Task Two – "A. Draw a parallelogram …" and turn the page to face the student

Say: Draw a parallelogram in this empty space here.

Once the student has finished drawing, then
1. Ask: What can say about the angles of this parallelogram?
   - If the student says, "the opposite angles are equal", or "all the vertex angles add up to 360°, or "the adjacent angles add up to 180°", then
      o Say: Write down your answer(s), and convince me.
   After the student has finished explaining his/her conclusion, then
      Ask: Is there any other relationship among the angles of this parallelogram?
   - If the student says, "all the vertex angles add up to 360°", then
      o Say: Write down your answer(s), and convince me.
   - If the student says, "no, that's all", then

2. Ask: What can you say about the sides of this parallelogram?
   - If the student says, " Opposite sides are equal", or "opposite sides are parallel", then
      o Say: Write down your answer(s) and convince me.
      After the student has finished explaining his/her conclusion, then
      Ask: Is there any other relationship involving the sides of this parallelogram?

Present Task Two – "B. Draw a new parallelogram …" and turn the page face to the student

Say: In the empty space here, draw a new parallelogram that is different from the one you drew previously.

Once the student finished drawing, then
1. Ask: Why is this a different parallelogram from the first one you drew?

2. Ask: What can you say about the angles of this parallelogram?
- If the student draws another parallelogram, then his/her answer to this question might be identical to Task Two A. No need to repeat the process as in Task Two A.
- If the student draws a rectangle, or a square, or a rhombus, and provides the same answer as he/she did in Task Two A., then
  - Say: Convince me.

3. Ask: What can you say about the sides of this parallelogram?
- If the student draws another parallelogram, then his/her answer to this question might be identical to Task Two A. If so, then ask question 4, "what can you say about the diagonals of this parallelogram?"
- If the student draws a rectangle, or a square, or a rhombus, and provides the same answer as he/she did in Task Two A., then Say: Convince me.

4. What can you say about the diagonals of this parallelogram?
- If the student draws a parallelogram, after she/he has finished describing the diagonals of the parallelogram,
  - Ask: Why?
    (Present a drawing of a rectangle), and then
  - Ask: What can you say about the diagonals of this one? Ask: Why?
    (Present a drawing of a square), and then
  - Ask: What can you say about the diagonals of this one?
        Ask: Why?
    (Present a drawing of a rhombus), and then
  - Ask: What can you say about the diagonals of this one?
  - Ask: Why?
- If the student draws a rectangle as a new parallelogram, after she/he has finished describing the diagonals of the rectangle,
  - Ask: Why?
    (Present a drawing of a square), and then
  - Ask: What can you say about the diagonals of this one?

- o Ask: Why?
  (Present a drawing of a rhombus), and then
- o Ask: What can you say about the diagonals of this one?
  - Ask: Why?
- If the student draws a square as a new parallelogram, after he/she has finished describing the diagonals of the square,
  - o Ask: Why?
    (Present a drawing of a rectangle), and then
  - o Ask: What can you say about the diagonals of this one?
  - o Ask: Why?
    (Present a drawing of a rhombus), and then
  - o Ask: What can you say about the diagonals of this one?
    Ask: Why?
- If the student draws a rhombus as a new parallelogram, after he/she has finished describing the diagonals of the rhombus,
  - o Ask: Why?
    (Present a drawing of a square), and then
  - o Ask: What can you say about the diagonals of this one?
  - o Ask: Why?
    (Present a drawing of a rectangle), and then
  - o Ask: What can you say about the diagonals of this one?
    Ask: Why?

5. Is it true that in every parallelogram the diagonals have the same midpoint (bisect each other)?

Ask: Why? Or Why not?

After the student has finished describing his/her conclusion, then
Say: write it down

Turn off the cameras and audio recorder. Remind the pair to write the date and their names on all the worksheets.

Say: I will collect all your worksheets.

About the Author

Sasha Wang is an assistant professor of mathematics education in the Department of Mathematics at Boise State University, Idaho, United States. She holds a M.S. in mathematics and a Ph.D. in mathematics education from Michigan State University. After graduate training in mathematics, she taught under-graduate mathematics for 10 years, and worked with K-12 teachers. She is interested in qualitative research methods, mathematical thinking and learning, and classroom discourse practices. Her research crosses disciplinary boundaries and is published in mathematics and science education, and curriculum studies

References

Ada, T., & Kurtuluş, A. (2010). Students' misconceptions and errors in transformation geometry, International Journal of Mathematical Education in Science and Technology, 41(7), 901-909.

Battista, M. T. (2007). The development of geometric and spatial thinking. In F. K. Lester (Ed.), *Second handbook of research on mathematics teaching and learning* (Vol. 2, pp. 483-908). Charlotte, NC: Information Age Publishing.

Battista, M. T. (2009). Highlights of research on learning school geometry. In T.V. Craine & R. Rubenstein (Eds.), *Understanding geometry for a changing world* (pp. 91-108). Reston, VA: National Council of Teachers of Mathematics.

Battista, M. T., & Clements, D. H. (1995). Geometry and proof. *The Mathematics Teacher*, 88(1), 48-54.

Banilower, E. R., Smith, P. S., Weiss, I. R., Malzahn, K. A., Campbell, K. M., & Weis, A. M. (2013). Report of the 2012 National Survey of Science and Mathematics Education. Chapel Hill, NC: Horizon Research, Inc.

Ball, D. L. (1993). With an eye on the mathematical horizon: Dilemmas of teaching elementary school mathematics. *The Elementary School Journal*, 93(4), 373-397.

Burger, W. F., & Shaughnessy, J. M. (1986). Characterizing the van Hiele levels of development in geometry. *Journal for Research in Mathematics Education*, 17(1), 31-48.

Clements, D. H. (1991). Elaborations on the levels of geometric thinking. Paper presented at the III International Symposium for Research in Mathematics Education, Valencia, Spain.

Clements, D. H. (2003). Teaching and learning geometry. In J. Kilpatrick (Ed.) *A research companion to principles and standards for school mathematics* (pp. 151-178). Reston, VA: National Council of Teachers of Mathematics.

Clements, D. H., & Battista, M. T. Geometry and spatial reasoning. (1992). In A.G. Grouws (Ed.), *Handbook of research on mathematics teaching and learning* (pp. 420-464). Reston, VA: National Council of Teachers of Mathematics.

Clements, D. H., Swaminathan, S., Hannibal, M. A. Z., & Sarama, J. (1999). Young children's concepts of shape. *Journal for Research in Mathematics Education, 30*, 192-212.

Crowley, M. L. (1987). The van Hiele model of the development of geometric thought. In M. Lindquist (Ed.), *Learning and teaching geometry, K-12* (pp. 1-16). Reston, VA: National Council of Teachers of Mathematics.

Crowley, M. L. (1990). Criterion referenced reliability indices associated with the van Hiele geometry test. *Journal for Research in Mathematics Education, 21*, 238-241.

Darken, B. (2007, May/June). Educating future elementary school and middle school teachers. *Focus,* 20-21.

De Villers, M. (1987). Research evidence on hierarchical thinking, teaching strategies and the van Hiele theory: Some critical comments. Stellenbosch, R. SouthAfrica: RUMEUS; Fac. Of Educ.; Univ. of Stellenbosch.

De Villers, M. (1999). The future of secondary school geoemtry. http://www.lettredlapreuve.it/Resume/deVilliers/deVilliers98/deVilliers982.html

Dingman, S., Teuscher, D., Newton, J. A., & Kasmer, L. (2013). Common mathematics standards in the United States: A comparison of K-8 state and Common Core standards

Floden, R. E. (2002). The measurement of opportunity to learn. In Board on International Comparative Studies in Education, A. C. Porter & A. Gamoran (Eds.), *Methodological advances in cross-national surveys of educational achievement (pp.231-266)*. Washington, DC: National Academy Press.

Fujita, T., & Jones, K. (2006). Primary trainee teachers' knowledge of parallelograms. Proceedings of the British Society for Research into Learning Mathematics, 26(2), 25-30.

Fuys, D., Geddes, D., & Tischler, R (1988). *The van Hiele model of thinking in geometry among adolescents.* (Journal for Research in Mathematics

Education Monograph No. 3). Reston, VA: National Council of Teachers of Mathematics.

Gutierrez, A., Jaime, A., & Fortuny, J. M. (1991). An alternative paradigm to evaluate the acquisition of the van Hiele levels. *Journal for Research in Mathematics Education, 22*, 237-251.

Gutierrez, A., Jaime, A. (1998). On the assessment of the van Hiele levels of reasoning. *Focus on Learning Problems in Mathematic, 20* (2&3): 27-46.

Hoffer, A. (1981). Geometry is more than proof. *Mathematics Teacher, 74*, 11-18.

Hoffer, A. (1983). van Hiele-based research. In R.Lesh & M. Landau (Eds.), *Acquisition of mathematics concepts and processes* (pp.205-227). New York, NY: Academic Press.

Jones, K. (2000). Providing a foundation for deductive reasoning: Students' interpretations when using dynamic geometry software and their evolving mathematical explanations. Educational Studies in Mathematics, 44(1–3), 55–85.

Jones, K., Mooney, C. and Harries, T. (2002). Trainee primary teachers' knowledge of geometry for teaching, Proceedings of the British Society for Research into Learning Mathematics, 22(1&2): 95–100.

Kerslake, D. (1991). The language of fractions. In K. Durkin & B. Shire (Ed.). Language in mathematical education: Research and practice. Chapter 8. Open Univesity Press: Bristol, PA. 85-94.

Lampert, M. (1990). When the problem is not the question and the answer is not the solution: Mathematical knowing and teaching. *American Educational Research Journal* 27 (1), 29- 63.

Lampert, M. (1998) Introduction. In *Talking mathematics in school*. M. Lampert & M. L. Blunk (Eds.). Cambridge University Press.

Li, W. (2013). *Secondary preservice teachers' mathematical discourses on geometric transformations*. (Doctoral dissertation). Retrieved from Indigo @ University of Illinois at Chicago.

Mayberry, J. (1983). The van Hiele levels of geometric thought in undergraduate preservice teachers. *Journal for Research in Mathematics Education, 14*(1), 58-69

Morgan, C. (1996). 'The language of mathematics': Towards a critical analysis of mathematics texts. *For the learning of Mathematics, 16 (3)*, 2-10.

Morgan, C. (1998). Writing mathematically: The discourse of investigation. London: Falmer Press.

Morgan, C. (2005). Words, definitions and concepts in discourses of mathematics, teaching and learning. *Language and Education. 19,*103-117.

Moschkovich, J. (2002). An introduction to examining everyday and academic mathematical practices. In M. Brenner & J. Moschkovvich (Eds), *Everyday and academic mathematics in the classroom.* JRME Monograph Number 11. (pp.1-11). Reston, VA: NCTM

Moschkovich, J. (2010). Language(s) and learning mathematics: Resources, challenges, and issues for research. In J. Moschkovich (Ed.), Language and mathematics education: multiple perspectives and directions for research (pp 1-28). Charlotte, NC: Information Age Publishing.

National Council of Teachers of Mathematics. (2000). *Principles and standards for school mathematics.* Reston, VA: National Council of Teachers of Mathematics.

National Council of Teachers of Mathematics (2003). *A research companion to principles and standards for school mathematics.* Reston, VA: National Council of teachers of Mathematics.

Newton, J. (2010). K-8 geometry state standards: A look through the lens of van Hiele's levels of geometric thinking. In J. P. Smith & J. E. Tarr (Eds.) The intended mathematics curriculum as represented in state-level curriculum standards: Consensus or confusion? (Vol. 2, pp. 59-87). Charlotte, NC: Information Age Publishing.

NGACBP (National Governors Association Center for Best Practices). 2010. Common Core State Standards for Mathematics. Available at www. corestandards.org/assets/CCSSI_ Math%20standards.pdf

Pimm, D. (1987). Speaking mathematically: Communication in mathematics classrooms. London: Routledge & Kegan Paul.

Rowland, T. (1995). Between the lines: The languages of mathematics. In J. Anghileri (Ed.), Children's Mathematical Thinking in the Primary Years (pp. 54-73). London: Cassell.

Sarama, J., & Clements, D. H. (2009). Early childhood mathematics education research: Learning trajectories for young children. New York: Routledge.

Schleppegrell, M.J. (2007). Language in mathematics teaching and learning: A research review. *Prepared for the Spencer Foundation.*

Senk, S. L. (1983). Proof-writing achievement and van Hiele levels among secondary school geometry students (Doctoral dissertation, The University of Chicago). *Dissertation Abstracts International, 44*, 417A.

Senk, S. L. (1989). van Hiele levels and achievement in writing geometry proofs. *Journal for Research in Mathematics Education, 20*(3), 309-321.

Sfard, A. (2000). On reform movement and the limits of mathematical discourse, *Mathematical Thinking and Learning, 2*(3). 157-189.

Sfard, A. (2005). Why cannot children see as the same what grown-ups cannot see as different/Early numerical thinking revisited. *Cognition and Instruction, 23*(2), 237-309.

Sfard, A. (2007). When rules of discourse change, but nobody tells you: making sense of mathematics learning from a commognitive standpoint. *The Journal of the Learning Science.* 16(40) 567-615.

Sfard, A. (2008). Thinking as communicating: human development, the growth of discourses, and mathematizing: Cambridge.

Shulman, L. S. (1986). Those who understand: Knowledge growth in teaching. Educational Researcher, 15(2), 4-14.

Smith, M. S. & Stein, M. K. (2011). 5 Practices for Orchestrating Productive Mathematics Discussions. Reston, Va., and Thousand Oaks, Calif.: National Council of Teachers of Mathematics and Corwin Press.

Thurston, W. (1994). On the proof and progress in mathematics. *American Mathematical Society,* 30 (2) 161-177.

Usiskin, Z. (1982). van Hiele levels and achievement in secondary school geometry (Final report of the Cognitive Development and Achievement in Secondary School Geometry Project: ERIC Document Reproduction Service No. ED 220 288). Chicago: University of Chicago.

Usiskin, Z. (1996). Mathematics as a language. In P. Elliott, & M. J. Kenney (Ed.), Communication in Mathematics, K-12 and beyond (pp. 231-243). National Council of Teachers of Mathematics.

Usiskin, Z., Griffin, J., Witonsky, D., & Willmore, E. (2008). The classification of quadrilaterals: A study of definition. Charlotte, NC: Information Age Publishing.

van Hiele, P. M. (1986). *Structure and insight: A theory of mathematics education.* Orlando, FL: Academic Press.

van Hiele, P. M. (1959/1985). The child's thought and geometry. In D. Fuys, D. Geddes & R. Tischler (Eds.), *English translation of selected writings of Dina van Hiele-Geldof and Pierre M. van Hiele* (pp. 243-252). Brooklyn, NY: Brooklyn College, School of Education.

van Hiele, P. M. (1999). Developing geometric thinking through activities that begin with play. *Teaching Children Mathematics, 5*(6), 310-315.

Vygotsky, L. S. (1997). The Collected Works of L.S. Vygosky (Vol. 4): *The History of the Development of Higher Mental Functions.* NY: Springer.

Wilson, M. (1990). Measuring a van Hiele geometry sequence: A reanalysis. *Journal for Research in Mathematics Education, 21,* 230-237.

Wilson, S. M., Floden, R. F., & Furrini-Mundy, J. (2001, February). *Teacher preparation research: Current knowledge, gaps, and recommendations.* Center for the Study of Teaching Policy, University of Washington, Seattle, WA.

Wirszup, I. (1976). Breakthroughs in the psychology of learning and teaching geometry. In J. L. Martin (Ed.), *Space and geometry: Papers from a research workshop* (pp. 75-97). Columbus, OH: ERIC/SMEAC.